Table of Symbols

Symbol	*Meaning*
$\in$	element of
$\notin$	not an element of
$\subseteq$	subset of
$\nsubseteq$	not a subset of
$\varnothing$	empty set
$\mathfrak{C}, \mathfrak{D}$	covers of sets
$\mathscr{C}$	open half-line topology for $\mathbb{R}$
$\mathscr{U}$	usual topology for $\mathbb{R}$
$\mathscr{H}$	half-open interval topology for $\mathbb{R}$
$\mathscr{I}$	indiscrete topology
$\mathscr{D}$	discrete topology
$U \cap V$	intersection of two sets
$U_1 \cap U_2 \cap \cdots \cap U_n$	intersection of a finite number of sets
$\bigcap\{U_\alpha : \alpha \in \Lambda\}$	intersection of an indexed collection of sets
$U \cup V$	union of two sets
$U_1 \cup U_2 \cup \cdots \cup U_n$	union of a finite number of sets
$\bigcup\{U_\alpha : \alpha \in \Lambda\}$	union of an indexed collection of sets
iff	if and only if
$\Rightarrow$	implies
$X - U$	complement of a set
$U - V$	difference of sets
(x, y)	ordered pair
$(x_1, x_2, \ldots, x_n)$	ordered n-tuple
$X \times Y$	product of two sets
$X_1 \times X_2 \times \cdots \times X_n$	product of a finite number of sets
$\times\{U_\alpha : \alpha \in \Lambda\}$	product of an indexed collection of sets
$f : X \to Y$	function from X to Y
$f(U)$	image of a set
$f^{-1}(U)$	inverse image of a set
$g \circ f$	composition of functions
$f\vert_U$	restriction of a function
id_X	identity function on X
$G(f)$	graph of a function
$>$	greater than
$\geqq$	greater than or equal to
$<$	less than
$\leqq$	less than or equal to

Symbol	*Meaning*
(a, b)	open interval
$[a, b]$	closed interval
$[a, b), (a, b]$	half-open interval
$(a, +\infty), (-\infty, a)$	open half-line
$[a, +\infty), (-\infty, a]$	closed half-line
$\mathrm{Cl}(U)$	closure of a set
$\mathrm{Int}(U)$	interior of a set
$\mathrm{Ext}(U)$	exterior of a set
$\mathrm{Bd}(U)$	boundary of a set
A'	set of all limit points of A
$\mathbb{R}$	real numbers
$\mathbb{R}^+$	positive real numbers
$\mathbb{Z}$	integers
$\mathbb{Z}^+$	positive integers
$\mathbb{Q}$	rational numbers
$\mathbb{R}^n$	n-dimensional Euclidean space
$d(x, y)$	distance between x and y
$B_r(y)$	open ball with center y and radius r
$B_r^-(y)$	closed ball with center y and radius r
■	end of proof

THE GREEK ALPHABET

Character	*Caps*	*Lowercase*	*Character*	*Caps*	*Lowercase*
Alpha	A	α	Nu	N	ν
Beta	B	β	Xi	Ξ	ξ
Gamma	Γ	γ	Omicron	O	o
Delta	Δ	δ	Pi	Π	π
Epsilon	E	ϵ	Rho	P	ρ
Zeta	Z	ζ	Sigma	Σ	σ
Eta	H	η	Tau	T	τ
Theta	Θ	θ	Upsilon	Υ	υ
Iota	I	ι	Phi	Φ	ϕ
Kappa	K	κ	Chi	X	χ
Lambda	Λ	λ	Psi	Ψ	ψ
Mu	M	μ	Omega	Ω	ω

Introduction to Topology

Introduction to Topology

Crump W. Baker
Indiana University Southeast

Krieger Publishing Company
Malabar, Florida

Original Edition 1991
Reprint Edition 1997 with corrections

Printed and Published by
KRIEGER PUBLISHING COMPANY
KRIEGER DRIVE
MALABAR, FLORIDA 32950

Library of Congress Cataloging-In-Publication Data

Baker, Crump W.
Introduction to topology / Crump W. Baker
p. cm.
Originally published: Dubuque, IA : Wm. C. Brown Publishers, c1991.
Includes bibliographical references (p. 151) and index.
ISBN 1-57524-008-4 (hardcover : alk. paper).
1. Topology. I. Title.
QA611.B25 1997
514'.3--dc20 96-26557
CIP

10 9 8 7 6 5 4 3

For Mildred M. Baker

Contents

*These sections are optional. Any material depending upon these sections is also marked with an asterisk (*).

Preface

This book is intended as a text for a one-semester undergraduate course in topology. It can also be used as an introduction to abstract mathematics. The book is accessible to students with only an elementary calculus background. In particular abstract algebra is not a prerequisite.

The fundamental concepts of general topology are covered rigorously but at a gentle pace and at an elementary level. Those aspects of general topology which depend only upon the elementary properties of sets and functions are emphasized. Due to the level of the text, there are some compromises. For example, the notion of a countable set is used sparingly and somewhat informally. Also a few theorems, such as the Tychonoff Theorem, are stated without proof. The motivation for this text lies in the belief that this part of general topology is accessible to students with only a calculus background and that this material provides an excellent introduction to abstract mathematics. It is the author's belief that the theory of elementary general topology is ideal for teaching students to prove theorems.

The exercises are an integral part of this text. Many routine exercises are included. Each exercise set begins with concrete, computational problems and then moves on to problems requiring simple proofs. The proofs of many of the theorems in the text are left as exercises. Each chapter ends with approximately twenty true-false questions. The student is required to explain each true statement and to find a counterexample for each false statement. In general the text contains a large number of problems which should be accessible to most students.

Elementary examples of topological spaces are emphasized. Many simple examples in which the student can easily calculate such objects as the interior or closure of a set are included. Specifically, the usual topology, half-open interval topology, and open half-line topology on $\mathbb{R}$ are emphasized throughout the text.

Venn diagrams and other types of illustrations are frequently used. Hopefully these diagrams clarify some of the more complicated concepts and proofs.

This book is intended for a traditional lecture style course in topology. However, many proofs and other details are omitted and may be supplied by the students in class or in written assignments. The instructor may choose to supply some proofs and assign the remaining ones.

Chapter 1 is central to this text. It is longer and perhaps more elementary than beginning chapters in most comparable texts. The elementary properties of sets and functions are covered in detail. As in the rest of the text, many of the proofs are left as exercises. Images and inverse images of sets are emphasized in preparation for the study of continuity in Chapters 2 and 3.

In Chapter 2 the general topological space, rather than the metric space, is developed. There are two reasons for this choice. First, the initial theorems are easier for the students to prove and simpler for general topological spaces and there are more elementary examples of general topological spaces than there are of metric spaces. Second, early development of the general topological space helps accomplish a secondary goal of the text, that of introducing the student to abstract mathematics. The usual related topics such as closure, interior, and limit point are developed in this chapter. Although continuity is not formally developed in this chapter, it is discussed and used as a motivation for the general topological space. Continuity is used as a link between elementary calculus and topology.

Subspaces, continuity, and homeomorphisms are covered in Chapter 3. Continuity is formally developed and several characterizations are given. After subspaces and homeomorphisms are discussed, a geometric or intuitive meaning of homeomorphic subspaces of Euclidean n-space is developed.

Chapter 4 covers the basic properties of product spaces for both finite and infinite products. The development of infinite products is in an optional section and any material elsewhere in the text that depends upon infinite products is marked with an asterisk (*). The development of the infinite product is slightly more difficult than the remainder of the text and, depending upon the background of the class, the instructor may want to omit this material. Very little of the remainder of the text uses the infinite product. The notion of a product space is used to construct a basis for Euclidean n-space (different from the basis developed in Chapter 3).

Chapters 5 and 6 cover the elementary properties of connected spaces and compact spaces, respectively. Connectedness is used to derive the Intermediate Value Theorem and some of its consequences. The Heine–Borel Theorem is proved. Compactness is characterized in terms of the finite intersection property, and Cantor's Theorem of Deduction is stated.

The major separation properties are covered in Chapter 7. This chapter is intended as a survey. Several of the more esoteric separation properties are briefly mentioned in order for the student to know that such properties exist. However, their development is left for a more advanced course.

Chapter 8 is a brief introduction to metric spaces. The concept of a metric space is introduced as a special case of the general topological space. The separation properties of metric spaces are investigated. Sequences and complete metric spaces are developed briefly.

Acknowledgments

I owe a special debt of gratitude to Professor James R. Boone of Texas A&M University. He reviewed several versions of this manuscript and made many valuable suggestions for its improvement. Section 4.4 was contributed by Professor Boone. I am deeply grateful to my wife, Dr. Alta M. Burnett, who encouraged and supported me during the writing of this text. She also proofread various drafts of the manuscript. I appreciate greatly the support and assistance given to me by my editor, Theresa Grutz, and the staff at William C. Brown and the many helpful suggestions made by the following reviewers: James R. Boone, Texas A&M University; William G. Fleissner, University of Kansas; Jim Henderson, Colorado College; Christopher McCord, University of Cincinnati; David Sutherland, Middle Tennessee State University; J. Paul Vicknair, California State University–San Bernardino.

I am grateful to Michael Riley for typing several sections of the manuscript. I had many valuable conversations concerning topology and word processing with my colleagues at Indiana University Southeast, principally with Professors Johnny Brown, Margaret Ehringer, Lawrence Mand, Timothy Miller, and James Woeppel. Finally, I appreciate the release time from teaching granted by Indiana University Southeast during part of the preparation of this manuscript.

New Albany, Indiana C. W. Baker

1

Preliminary Topics

1.1 *Topology*

One of the origins of the subject matter which has become known as topology lies in early work done on the extension and generalization of the concept of continuity. This work was done in the late 1800s and early 1900s primarily by Maurice Frechet (1878–1973) and Felix Hausdorff (1868–1942). Continuity, you will recall, is a central part of the calculus. Essentially every major theorem in calculus involves the concept of continuity.

The reason for extending and generalizing continuity lies in the exceptional properties of continuity. For example, the Intermediate Value Theorem states that if a function f is continuous on a closed interval $[a, b]$, then the function assumes all values between $f(a)$ and $f(b)$. In other words, the continuous image of an interval is still an interval in the sense that there are no holes or breaks in the image. (For the purpose of this discussion we shall agree that a single point is a very short interval.) It is easily seen for an increasing function that if there were a break or hole in the image, then there would be a value y_0 between $f(a)$ and $f(b)$ that would not be assumed by the function. (See Figure 1.1.1.) The continuous image of an interval may be longer or shorter than the original interval, but it will still be an interval. In Figure 1.1.2 the function f given by $f(x) = 2x$ stretches the interval $[1, 2]$ by a factor of 2. Observe that in Figure 1.1.3 the function f defined by $f(x) = x/2$ shrinks the interval $[2, 4]$ by a factor of 1/2. However, in both cases the image of the interval is still an interval. This notion of stretching or shrinking but not breaking a geometric object such as a line segment is a central concept of topology.

This extension and generalization of continuity involved extending continuity to functions between sets of objects other than real numbers. Sets

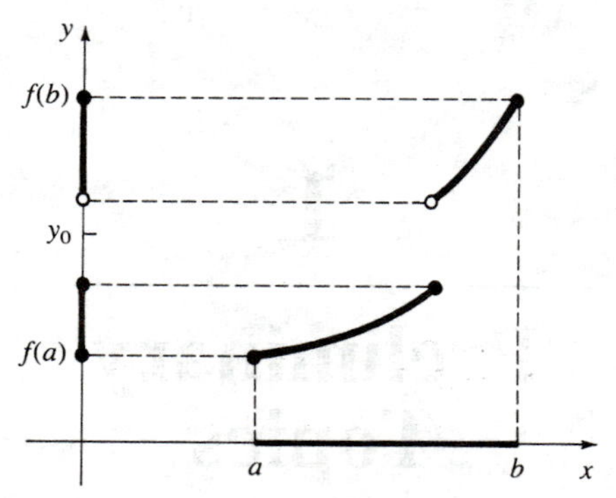

Figure 1.1.1

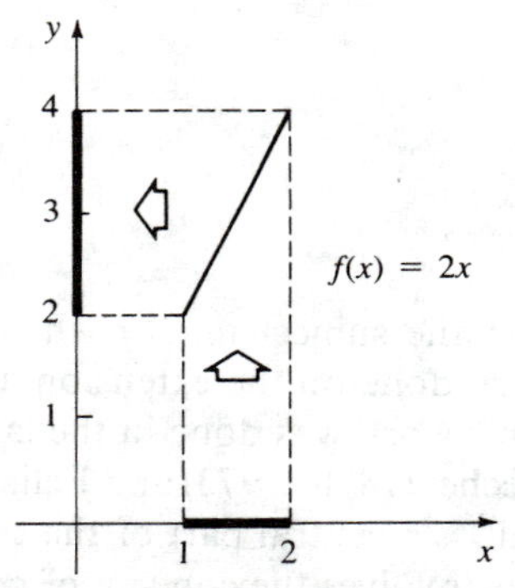

Figure 1.1.2

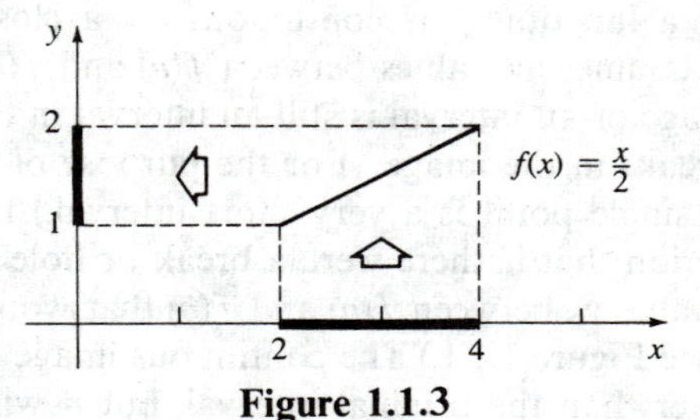

Figure 1.1.3

of objects such as ordered pairs of real numbers, matrices, and functions were investigated. Eventually continuity was extended to functions between "abstract" sets whose elements were fixed but not specified. This extension was made possible by extensive work done on the theory of sets by Georg Cantor (1845–1918) in the 1880s. Two methods were found for extending the notion of continuity. The first method, found by Frechet in 1906, consisted of extending the notion of distance. A "distance function" with properties similar to those of distance on the real number line was defined for an arbitrary set X.

Continuity was then defined in terms of this distance function. The second method, developed by Hausdorff in 1914, involved the generalization of the concept of an open interval. Certain subsets of an arbitrary set X with properties similar to those of open intervals were designated as "open sets" and used to define continuity.

A function is continuous provided that points in the domain that are near one another are mapped to points in the range that are near one another. That is, a function f is continuous if for each point x_0 in the domain, whenever x is near x_0 then $f(x)$ is near $f(x_0)$. Thus a notion of "nearness" or "closeness" is required in order to extend and generalize the concept of continuity. If the concept of a distance function is used, two points are considered to be near one another if the distance between the points is small. On the other hand, if the notion of an open set is used, two points are considered to be near one another if the points are in the same open set.

A set with the additional structure involving the notion of distance is called a metric space, and a set with the additional structure involving open sets is referred to as a topological space. The term "space" is frequently used in mathematics to refer to a set with some structure added. For example, if you have studied linear algebra, you are familiar with vector spaces which consist of a set along with the operations of vector addition, scalar multiplication, and the appropriate axioms.

Although motivated by the desire to extend the concept of continuity, the notions of a topological space and a metric space have become themselves major objects of study in mathematics. This text is primarily concerned with the basic properties of topological spaces and metric spaces. Although historically the concept of a metric space was developed before that of a topological space, we shall start with the topological space. It is a much simpler structure than the metric space, and the proofs of the basic properties are less complicated. The theory of topological spaces is developed entirely in terms of sets and functions. The remaining sections of this chapter cover the basic properties of sets and functions required for the study of topology.

1.2 *Sets*

Intuitively a set is a collection of objects. If x is an object and A is a set, we write "$x \in A$" to denote that x is a member of A or "$x \notin A$" to denote that x is not a member of A. For example, if $A = \{-1, 0, 2\}$, then $-1 \in A$ and $3 \notin A$. Note that we simply listed the elements of the set A. For some sets the following notation is more convenient than listing the elements:

$$A = \{x : P(x)\}$$

where $P(x)$ is a property that x satisfies exactly when x is a member of the set A. This notation is read as follows: "A is the set of all x such that the property $P(x)$ holds." The colon is read as "such that." (Some authors use a vertical line in place of the colon.)

Example 1.2.1

Let $A = \{x : x \text{ is an integer and } x > 3\}$. That is, A is the set of all integers greater than 3. The statement "x is an integer and $x > 3$" is the property $P(x)$. Another way of describing A is $A = \{4, 5, 6, \ldots\}$.

There are several special sets of numbers that will be used throughout this text. The following notation will be used:

$\mathbb{Z}$ is the set of integers.

$\mathbb{Z}^+$ is the set of positive integers.

$\mathbb{Q}$ is the set of rational numbers (that is, all real numbers that can be expressed in the form a/b where a and b are integers and $b \neq 0$).

$\mathbb{R}$ is the set of real numbers.

$\mathbb{R}^+$ is the set of positive real numbers.

In practice a slightly abbreviated form of notation is used for sets. The set in Example 1.2.1, $A = \{x : x \text{ is an integer and } x > 3\}$, is usually expressed as $A = \{x \in \mathbb{Z}^+ : x > 3\}$.

The following notation will be used for intervals on the real number line. Assume that $a, b \in \mathbb{R}$ and that $a < b$:

$(a, b) = \{x \in \mathbb{R} : a < x < b\}$ is the *open interval* from a to b.

$[a, b] = \{x \in \mathbb{R} : a \leqq x \leqq b\}$ is the *closed interval* from a to b.

$[a, b) = \{x \in \mathbb{R} : a \leqq x < b\}$ is a *half-open interval* from a to b (open at the right endpoint and closed at the left endpoint).

$(a, b] = \{x \in \mathbb{R} : a < x \leqq b\}$ is a *half-open interval* from a to b (open at the left endpoint and closed at the right endpoint).

$(a, +\infty) = \{x \in \mathbb{R} : x > a\}$ is an *open half-line*.

$(-\infty, a) = \{x \in \mathbb{R} : x < a\}$ is an *open half-line*.

$[a, +\infty) = \{x \in \mathbb{R} : x \geqq a\}$ is a *closed half-line*.

$(-\infty, a] = \{x \in \mathbb{R} : x \leqq a\}$ is a *closed half-line*.

Definition 1.2.2

A set A is said to be a *subset* of a set B, denoted by $A \subseteq B$, provided that every element of A is also an element of B. If A is not a subset of B, we write $A \nsubseteq B$.

Obviously every set is a subset of itself. The set with no elements is called the *empty set* or *null set* and is denoted by $\emptyset$. The empty set is a subset of every set. To see this, let A be a set; suppose that the empty set is not a subset of A. Then there must exist an element x in the empty set that is not in A. This is obviously a contradiction and hence $\emptyset \subseteq A$.

Example 1.2.3

Let $A = \{1, 2, 3, 4\}$, $B = \{2, 4\}$, and $C = \{1, 3, 4\}$. Then $B \subseteq A$ and $C \subseteq A$. However $C \nsubseteq B$ and $B \nsubseteq C$.

Two sets are equal if the sets have exactly the same elements. The following statement defines equality of sets in terms of the subset relationship. This definition is convenient to use in proofs.

Definition 1.2.4

Sets A and B are said to be *equal* provided that $A \subseteq B$ and $B \subseteq A$.

Definition 1.2.5

If A and B are sets with $A \subseteq B$ and $A \neq B$, then A is called a *proper subset* of B.

Some authors make a distinction in notation for proper subsets writing "$A \subset B$" to denote that A is a proper subset of B. We shall use the notation "$A \subseteq B$" for all subsets.

There are two important operations on sets: the union of sets and the intersection of sets. By union of sets we simply mean the joining of the two sets together to form one set. The intersection of two sets is a set consisting of elements that the two sets have in common (see Figure 1.2.1). The following definition makes these ideas precise.

Definition 1.2.6

Let A and B be sets. The *union* of A and B is the set $A \cup B = \{x : x \in A \text{ or } x \in B\}$. The *intersection* of A and B is the set $A \cap B = \{x : x \in A \text{ and } x \in B\}$.

Definition 1.2.7

Two sets A and B are said to be *disjoint* provided that $A \cap B = \varnothing$.

Another important operation on sets is the "difference" of sets. In this operation one set is "subtracted" from another set.

Definition 1.2.8

The *difference* of sets A and B is the set $A - B = \{x : x \in A \text{ and } x \notin B\}$.

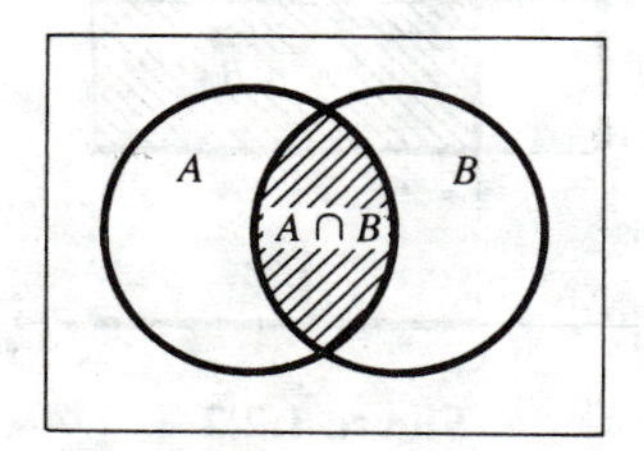

Figure 1.2.1

Note that B does not have to be a subset of A in order to form the difference $A - B$. If A and B are disjoint, $A - B = A$. Obviously if $A \subseteq B$, then $A - B = \emptyset$.

We usually assume that all sets under consideration are subsets of a fixed set X. In this case $X - A$ is called the *complement* of A.

Example 1.2.9

Let $A = [0,4]$, $B = (1,5]$, $C = (3,6)$, and $D = (1,+\infty)$. Then $B - A = (4,5]$, $A - B = [0,1]$, $C - D = \emptyset$, and $D - C = (1,3] \cup [6,+\infty)$.

The next operation we consider is the product of two sets. If x and y are elements, the ordered pair x and y is denoted by (x, y).

Definition 1.2.10

The *product* (or *Cartesian product*) of sets A and B is the set $A \times B = \{(x, y) : x \in A \text{ and } y \in B\}$.

The notation (x, y) for an ordered pair is the same as the notation for an open interval. However, it should be clear from the context which concept is indicated by the notation.

If A and B are intervals on the real number line with A on the x-axis and B on the y-axis, then $A \times B$ is a rectangular region in the xy-plane (see Figure 1.2.2).

Example 1.2.11

Let $A = \{1,2\}$ and $B = \{3,4,5\}$. Then

$$A \times B = \{(1,3), (1,4), (1,5), (2,3), (2,4), (2,5)\}.$$

Obviously if A and B are finite sets containing m elements and n elements, respectively, then $A \times B$ contains mn elements.

Example 1.2.12

If $A = [1,3)$ and $B = [1,2)$, then $A \times B$ is a "half-open" rectangular region in the xy-plane (see Figure 1.2.3).

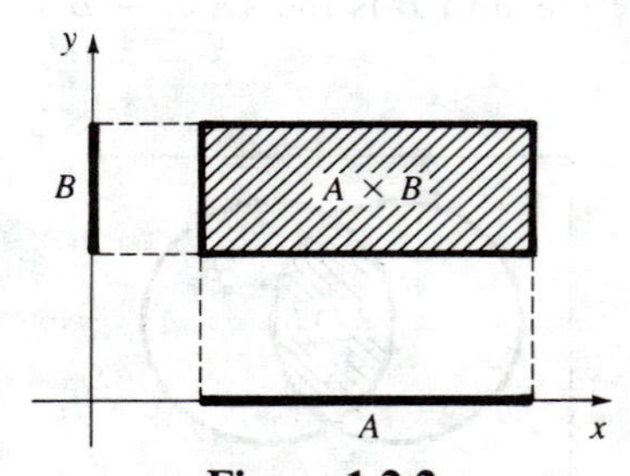

Figure 1.2.2

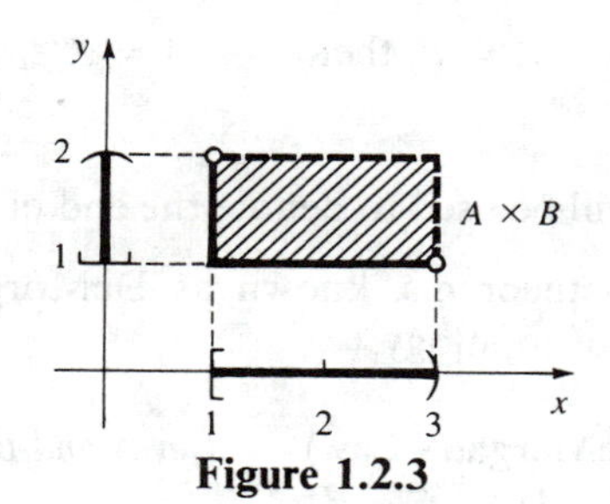

Figure 1.2.3

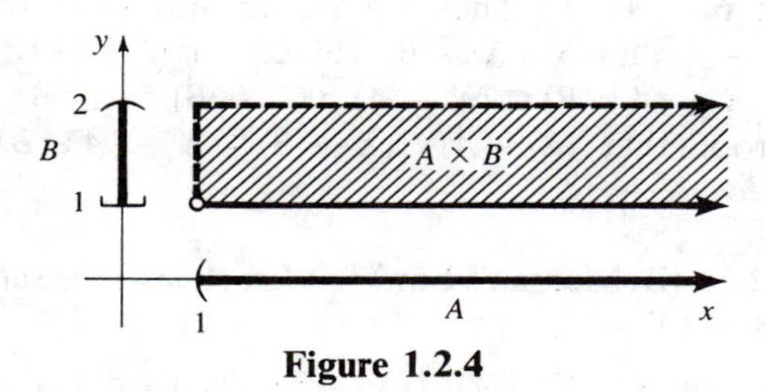

Figure 1.2.4

Example 1.2.13

Let $A = (1, +\infty)$ and $B = [1, 2)$. Then $A \times B$ is the rectangular strip in the xy-plane shown in Figure 1.2.4.

Most of the concepts in this book are explained and defined in terms of sets and the operations union, intersection, difference, and product. It is extremely important that you understand these operations on two levels. First you must have a firm intuitive idea of the meanings of these operations. For example, you must be able to picture the intersection of two sets as the "overlapping" of the sets, that is, the elements that the sets have in common. Second you must be able to understand and use the formal definitions of these operations in proofs.

The remainder of this section consists of theorems involving various properties of sets. These results form the basis for much of the material in the text.

The abbreviation "iff" will be used for the expression "if and only if" and arrows ($\Rightarrow$ or $\Leftarrow$) will be used to indicate the direction of a proof.

THEOREM 1.2.14 *Let A and B be subsets of X. Then $X - A = X - B$ iff $A = B$.*

Proof ($\Rightarrow$) Assume $X - A = X - B$. In order to show that $A \subseteq B$, suppose $x \in A$. Then $x \notin X - A$. Since $X - A = X - B$, $x \notin X - B$. Thus $x \in B$ and hence $A \subseteq B$. To see that $B \subseteq A$, suppose $x \in B$. Then $x \notin X - B$. Since $X - B = X - A$, $x \notin X - A$. Therefore $x \in A$ and hence $B \subseteq A$. Since $A \subseteq B$ and $B \subseteq A$, it follows that $A = B$.

The proof that if $A = B$, then $X - A = X - B$ is analogous and is left as an exercise. ■

The symbol "■" will be used to denote the end of a proof.

The following two theorems, known as DeMorgan's Laws, are used frequently in the study of topology.

THEOREM 1.2.15 (DeMorgan's Law) *Let A and B be subsets of X. Then* $X - (A \cup B) = (X - A) \cap (X - B)$.

Proof Let $x \in X - (A \cup B)$. Then $x \notin A \cup B$. Thus $x \notin A$ and $x \notin B$. Therefore $x \in X - A$ and $x \in X - B$. Hence $x \in (X - A) \cap (X - B)$. This proves that $X - (A \cup B) \subseteq (X - A) \cap (X - B)$.

The proof that $(X - A) \cap (X - B) \subseteq X - (A \cup B)$ is left as an exercise. ■

THEOREM 1.2.16 (DeMorgan's Law) *Let A and B be subsets of X. Then* $X - (A \cap B) = (X - A) \cup (X - B)$.

Proof Assume that $x \in (X - A) \cup (X - B)$. Then $x \in X - A$ or $x \in X - B$. Suppose $x \in X - A$. Then $x \notin A$. Therefore $x \notin A \cap B$ and hence $x \in X - (A \cap B)$. Similarly, if $x \in X - B$, then $x \notin B$. Thus $x \notin A \cap B$ which implies that $x \in X - (A \cap B)$. So in either case we have that $x \in X - (A \cap B)$. This proves that $(X - A) \cup (X - B) \subseteq X - (A \cap B)$.

The proof that $X - (A \cap B) \subseteq (X - A) \cup (X - B)$ is left as an exercise. ■

DeMorgan's Laws state that the complement of the intersection of two sets is the union of the complements of the sets and that the complement of the union of two sets is the intersection of the complements of the sets. In the next section we shall see that DeMorgan's Laws extend to arbitrary collections of sets.

The next two theorems deal with algebraic properties of sets. The first theorem states that intersection distributes over union, and the second states that union distributes over intersection.

THEOREM 1.2.17 *Let A, B, and C be sets. Then* $A \cap (B \cup C) = (A \cap B) \cup (A \cap C)$.

Proof Let $x \in A \cap (B \cup C)$. Then $x \in A$ and $x \in B \cup C$. Therefore $x \in A$ and either $x \in B$ or $x \in C$. Assume $x \in A$ and $x \in B$. Then $x \in A \cap B$ and hence $x \in (A \cap B) \cup (A \cap C)$. Similarly, if $x \in A$ and $x \in C$, then $x \in A \cap C$ and thus $x \in (A \cap B) \cup (A \cap C)$. It follows that $A \cap (B \cup C) \subseteq (A \cap B) \cup (A \cap C)$.

The proof that $(A \cap B) \cup (A \cap C) \subseteq A \cap (B \cup C)$ is left as an exercise. ■

THEOREM 1.2.18 *Let A, B, and C be sets. Then* $A \cup (B \cap C) = (A \cup B) \cap (A \cup C)$.

Proof Assume $x \in (A \cup B) \cap (A \cup C)$. Then $x \in A \cup B$ and $x \in A \cup C$. That is, $x \in A$ or $x \in B$, and $x \in A$ or $x \in C$. Now either $x \in A$ or $x \notin A$. If $x \in A$, then $x \in A \cup (B \cap C)$. If $x \notin A$, then $x \in B$ and $x \in C$. Therefore $x \in B \cap C$ and hence $x \in A \cup (B \cap C)$. Thus in either case $x \in A \cup (B \cap C)$. This proves that $(A \cup B) \cap (A \cup C) \subseteq A \cup (B \cap C)$.

The proof that $A \cup (B \cap C) \subseteq (A \cup B) \cap (A \cup C)$ is left as an exercise. ■

THEOREM 1.2.19 *Let A, B, and C be sets. If $A \subseteq B$, then $A \cup C \subseteq B \cup C$.*

Proof Assume $A \subseteq B$. Let $x \in A \cup C$. Then $x \in A$ or $x \in C$. Suppose $x \in A$. Since $A \subseteq B$, $x \in B$. Hence $x \in B \cup C$. If $x \in C$, then clearly $x \in B \cup C$. Thus in either case $x \in B \cup C$. Therefore $A \cup C \subseteq B \cup C$. ■

The proof of the following theorem is analogous to that of Theorem 1.2.19 and is left as an exercise.

THEOREM 1.2.20 *If A, B, and C are sets, then $A \subseteq B$ implies that $A \cap C \subseteq B \cap C$.*

The next theorem states that two sets are disjoint if and only if either set is contained in the complement of the other (see Figure 1.2.5).

THEOREM 1.2.21 *Let A and B be subsets of X. Then $A \cap B = \varnothing$* iff *$A \subseteq X - B$.*

Proof ($\Rightarrow$) Assume that $A \cap B = \varnothing$. Suppose $x \in A$. Since $A \cap B = \varnothing$, $x \notin B$. Thus $x \in X - B$. Therefore $A \subseteq X - B$.

The proof that if $A \subseteq X - B$, then $A \cap B = \varnothing$ is left as an exercise. ■

This section is concluded with several results concerning products of sets.

THEOREM 1.2.22 *If A and B are sets then $A \times B = \varnothing$* iff *$A = \varnothing$ or $B = \varnothing$.*

Proof ($\Rightarrow$) Assume $A \times B = \varnothing$. Suppose $A \neq \varnothing$ and $B \neq \varnothing$. Then there exist $x \in A$ and $y \in B$. Therefore $(x, y) \in A \times B$ which contradicts the assumption that $A \times B = \varnothing$. Thus $A = \varnothing$ or $B = \varnothing$.

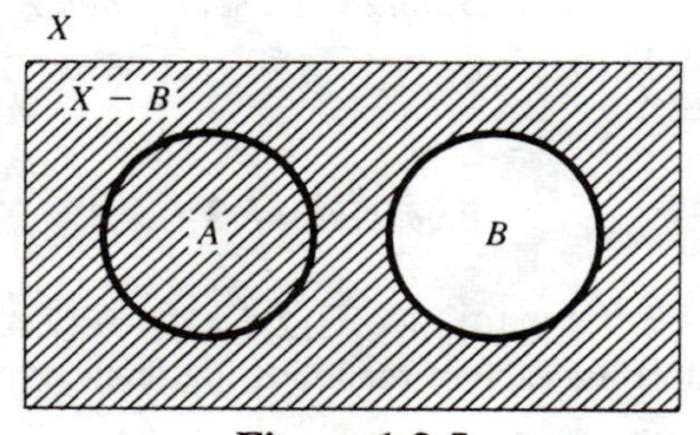

Figure 1.2.5

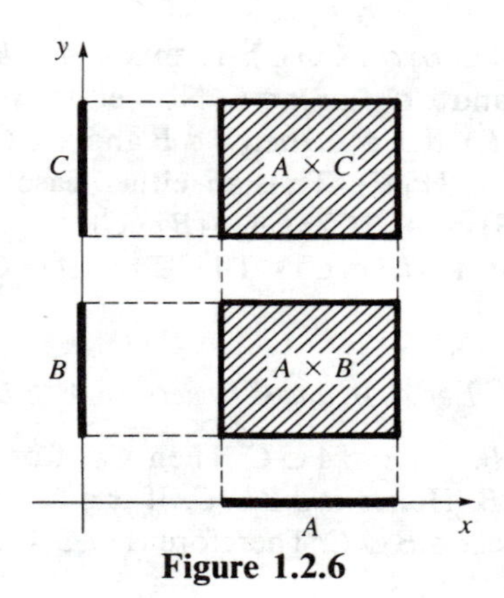

Figure 1.2.6

($\Leftarrow$) Assume either $A = \varnothing$ or $B = \varnothing$. Suppose $A \times B \neq \varnothing$. Then there is an ordered pair (x, y) in $A \times B$. Thus $x \in A$ and $y \in B$. Therefore $A \neq \varnothing$ and $B \neq \varnothing$ which is a contradiction. Therefore $A \times B = \varnothing$. ∎

The next theorem states that the product distributes over the intersection of sets.

THEOREM 1.2.23 *Let A, B, and C be sets. Then $A \times (B \cap C) = (A \times B) \cap (A \times C)$.*

Proof Suppose $(x, y) \in A \times (B \cap C)$. Then $x \in A$ and $y \in B \cap C$. Thus $x \in A$, $y \in B$, and $y \in C$. Hence $(x, y) \in A \times B$ and $(x, y) \in A \times C$. Therefore, $(x, y) \in (A \times B) \cap (A \times C)$. This proves that $A \times (B \cap C) \subseteq (A \times B) \cap (A \times C)$.

Suppose $(x, y) \in (A \times B) \cap (A \times C)$. Then $(x, y) \in A \times B$ and $(x, y) \in A \times C$. Then $x \in A$, $y \in B$, and $y \in C$. Therefore $x \in A$ and $y \in B \cap C$. Thus $(x, y) \in A \times (B \cap C)$. Hence $(A \times B) \cap (A \times C) \subseteq A \times (B \cap C)$.

This proves that $(A \times B) \cap (A \times C) = A \times (B \cap C)$. ∎

The next theorem of this section states that the product distributes over the union of sets (see Figure 1.2.6). The proof is similar to that of Theorem 1.2.23 and is left as an exercise.

THEOREM 1.2.24 *Let A, B, and C be sets. Then $A \times (B \cup C) = (A \times B) \cup (A \times C)$.*

Exercises 1.2

1. Let $X = \{1, 2, 3, 4, 5, 6, 7, 8, 9, 10\}$, $A = \{1, 2, 3, 4\}$, $B = \{2, 4, 6, 8\}$, $C = \{1, 3, 5, 7, 9\}$, and $D = \{3, 6, 9, 10\}$. Find each of the following sets:

   (a) $A \cup B$ **(b)** $X - (A \cup B)$ **(c)** $C \cap D$

(d) $(X - A) \cap C$ (e) $(A \cap B) \cap C$ (f) $A \cup (C \cap D)$
(g) $(A \cup C) \cap (A \cup D)$ (h) $(X - A) \cap (X - B)$ (i) $(A \cup B) \cap D$

2. Let $X = \mathbb{R}$, $A = [0, 5)$, $B = (3, 6]$, $C = (6, +\infty)$, and $D = [3, 5]$. Find each of the following sets:

(a) $A - B$ (b) $A \cup C$ (c) $B \cap C$
(d) $(A \cup B) - D$ (e) $X - A$ (f) $X - C$
(g) $(A \cap B) \cap D$ (h) $(A \cup B) \cap D$ (i) $B - A$

3. Let $A = [0, 2]$, $B = [3, 4)$, $C = [2, +\infty)$, $D = (-\infty, 5)$, and $E = (0, 3)$. Sketch each of the following sets in the xy-plane:

(a) $A \times A$ (b) $A \times B$
(c) $A \times C$ (d) $E \times B$
(e) $E \times (A \cup B)$ (f) $A \times (C \cap D)$

4. Let $A = [1, 3]$, $B = [1, 4]$, $C = [2, 4]$, and $D = [2, 5]$. Sketch each of the following sets in the xy-plane:

(a) $(A \times B) \cap (C \times D)$ (b) $(A \times B) \cup (C \times D)$
(c) $(A \cap C) \times (B \cap D)$ (d) $(A \cup C) \times (B \cup D)$.

5. Give an example of a subset of $\mathbb{R} \times \mathbb{R}$ that is not of the form $A \times B$, where $A \subseteq \mathbb{R}$, and $B \subseteq \mathbb{R}$. Sketch the set in the xy-plane.

6. Complete the proof of Theorem 1.2.14.

7. Complete the proof of Theorem 1.2.15 (DeMorgan's Law).

8. Complete the proof of Theorem 1.2.16 (DeMorgan's Law).

9. Give examples of sets to show that each of the following statements is *false*:

(a) $A - B = B - A$
(b) $X - (A \cup B) = (X - A) \cup (X - B)$
(c) $X - (A \cap B) = (X - A) \cap (X - B)$
(d) If $A \cup C \subseteq B \cup C$, then $A \subseteq B$.
(e) If $A \cap C \subseteq B \cap C$, then $A \subseteq B$.
(f) If $A \subseteq X - B$, then $X - A \subseteq B$.
(g) $(A \cup B) - C = A \cup (B - C)$

10. Complete the proof of Theorem 1.2.17.

11. Complete the proof of Theorem 1.2.18.

12. Prove Theorem 1.2.20.

13. Complete the proof of Theorem 1.2.21.

14. Let A and B be subsets of X. Prove that $A \subseteq X - B$ iff $B \subseteq X - A$.

15. Prove Theorem 1.2.24.

16. Let A, B, and C be sets. Prove that $(A - B) - C = (A - C) - (B - C)$.

1.3 *Extended Set Operations*

In this section the operations union, intersection, and product are extended to more general collections of sets. The product operation is extended to finite collections of sets, and the union and intersection operations are extended to infinite collections of sets.

Definition 1.3.1

Let $A_1, A_2, \ldots, A_n$ be a collection of n sets. The *union* of $A_1, A_2, \ldots, A_n$ is the set $A_1 \cup A_2 \cup \cdots \cup A_n = \{x : x \in A_i \text{ for some } i \in \{1, 2, \ldots, n\}\}$. The *intersection* of $A_1, A_2, \ldots, A_n$ is the set $A_1 \cap A_2 \cap \cdots \cap A_n = \{x : x \in A_i$ for every $i \in \{1, 2, \ldots, n\}\}$.

Example 1.3.2

Let $A_1 = [0, 4)$, $A_2 = (-1, 6)$, $A_3 = [3, 5]$, and $A_4 = [2, +\infty)$. Then $A_1 \cup A_2 \cup A_3 \cup A_4 = (-1, +\infty)$ and $A_1 \cap A_2 \cap A_3 \cap A_4 = [3, 4)$.

Definition 1.3.3

Let $A_1, A_2, \ldots, A_n$ be a collection of n sets. The product of $A_1, A_2, \ldots, A_n$ is the set $A_1 \times A_2 \times \cdots \times A_n = \{(x_1, x_2, \ldots, x_n) : x_i \in A_i \text{ for every } i \in \{1, 2, \ldots, n\}\}$. The element $(x_1, x_2, \ldots, x_n)$ is called an *ordered n-tuple*. It is the obvious generalization of the concept of an ordered pair.

Example 1.3.4

Let $A_1 = \{1, 2\}$, $A_2 = \{2, 3\}$, and $A_3 = \{4\}$. Then $A_1 \times A_2 \times A_3 = \{(1, 2, 4), (1, 3, 4), (2, 2, 4), (2, 3, 4)\}$.

A different notation is sometimes used for the union, intersection, and product of sets. The union of sets $A_1, A_2, \ldots, A_n$ can be denoted by $\bigcup\{A_i : i = 1, 2, \ldots, n\}$. Similarly the intersection of $A_1, A_2, \ldots, A_n$ is denoted by $\bigcap\{A_i : i = 1, 2, \ldots, n\}$, and the product of $A_1, A_2, \ldots, A_n$ is denoted by $\mathbf{X}\{A_i : i = 1, 2, \ldots, n\}$.

This new notation is easily extended to more general collections of sets. In the above notation, if we let $\Lambda = \{1, 2, \ldots, n\}$, then we have the union of the sets $A_1, A_2, \ldots, A_n$ denoted by $\bigcup\{A_i : i \in \Lambda\}$. Similarly the intersection of $A_1, A_2, \ldots, A_n$ is denoted by $\bigcap\{A_i : i \in \Lambda\}$.

Now there is no particular reason for the set Λ to consist of the numbers one through n or even to be finite. If Λ is any nonempty set and for each $\alpha \in \Lambda$ there is a set A_α, we can form the sets $\bigcup\{A_\alpha : \alpha \in \Lambda\}$ and $\bigcap\{A_\alpha : \alpha \in \Lambda\}$.

Before proceeding further some formal definitions are needed.

Definition 1.3.5

Let Λ be a nonempty set. If for each $\alpha \in \Lambda$ there is a set A_α, the collection $\{A_\alpha : \alpha \in \Lambda\}$ is called an *indexed collection* of sets. The set Λ is called the *index set*.

Example 1.3.6

Let $\Lambda = \mathbb{R}$ and for each $\alpha \in \Lambda$ let $A_\alpha = [\alpha, \alpha + 1)$. Then $\{A_\alpha : \alpha \in \Lambda\}$ is an indexed collection of sets.

Definition 1.3.7

Let $\{A_\alpha : \alpha \in \Lambda\}$ be an indexed collection of sets. The *union* of the collection $\{A_\alpha : \alpha \in \Lambda\}$ is the set $\bigcup\{A_\alpha : \alpha \in \Lambda\} = \{x : x \in A_\alpha \text{ for some } \alpha \in \Lambda\}$. The intersection of the collection $\{A_\alpha : \alpha \in \Lambda\}$ is the set $\bigcap\{A_\alpha : \alpha \in \Lambda\} = \{x : x \in A_\alpha \text{ for every } \alpha \in \Lambda\}$.

Example 1.3.8

Let $\Lambda = \mathbb{R}^+$ and for each $\alpha \in \Lambda$ let $A_\alpha = [0, 1 + \alpha)$. Then $\bigcap\{A_\alpha : \alpha \in \Lambda\} = [0, 1]$ and $\bigcup\{A_\alpha : \alpha \in \Lambda\} = [0, +\infty)$.

Example 1.3.9

Let $\Lambda = \mathbb{Z}^+$ and for each $\alpha \in \Lambda$ let $B_\alpha = (-\alpha, \alpha)$. Then $\bigcap\{B_\alpha : \alpha \in \Lambda\} = (-1, 1)$ and $\bigcup\{B_\alpha : \alpha \in \Lambda\} = \mathbb{R}$.

Example 1.3.10

Let $\Lambda = \mathbb{Z}$ and for each $\alpha \in \Lambda$ let $D_\alpha = [\alpha, \alpha + 1]$. Then $\bigcap\{D_\alpha : \alpha \in \Lambda\} = \varnothing$ and $\bigcup\{D_\alpha : \alpha \in \Lambda\} = \mathbb{R}$.

The product of sets also extends to indexed collections of sets. However, the process is somewhat complicated and will be developed later in the text.

A slightly different notation is sometimes used for collections of sets. If $\mathscr{A}$ is a (nonempty) collection of sets (sometimes called a *family* of sets), then the union of the sets in $\mathscr{A}$ is the set $\bigcup\{A : A \in \mathscr{A}\} = \{x : x \in A \text{ for some } A \in \mathscr{A}\}$. Similarly the intersection of the sets in $\mathscr{A}$ is the set $\bigcap\{A : A \in \mathscr{A}\} = \{x : x \in A \text{ for every } A \in \mathscr{A}\}$.

Many of the theorems in the previous section extend to indexed collections of sets. In particular, DeMorgan's Laws extend to indexed collections of sets.

THEOREM 1.3.11 (DeMorgan's Law for Indexed Sets) *Let $\{A_\alpha : \alpha \in \Lambda\}$ be an indexed collection of subsets of a set X. Then $X - \bigcup\{A_\alpha : \alpha \in \Lambda\} = \bigcap\{X - A_\alpha : \alpha \in \Lambda\}$.*

Proof Let $x \in X$. Then $x \in X - \bigcup\{A_\alpha : \alpha \in \Lambda\}$ iff $x \notin \bigcup\{A_\alpha : \alpha \in \Lambda\}$ iff $x \notin A_\alpha$ for every $\alpha \in \Lambda$ iff $x \in X - A_\alpha$ for every $\alpha \in \Lambda$ iff $x \in \bigcap\{X - A_\alpha : \alpha \in \Lambda\}$. Therefore $X - \bigcup\{A_\alpha : \alpha \in \Lambda\} = \bigcap\{X - A_\alpha : \alpha \in \Lambda\}$. ■

THEOREM 1.3.12 (DeMorgan's Law for Indexed Sets) *Let $\{A_\alpha : \alpha \in \Lambda\}$ be an indexed collection of subsets of a set X. Then $X - \bigcap\{A_\alpha : \alpha \in \Lambda\} = \bigcup\{X - A_\alpha : \alpha \in \Lambda\}$.*

The proof is analogous to that of Theorem 1.3.11 and is left as an exercise.

Theorems 1.2.17 and 1.2.18, which state that intersection distributes over union and that union distributes over intersection, respectively, also extend to indexed collections of sets.

THEOREM 1.3.13 *Let $\{A_\alpha : \alpha \in \Lambda\}$ be an indexed collection of sets and let B be a set. Then $B \cap (\bigcup\{A_\alpha : \alpha \in \Lambda\}) = \bigcup\{B \cap A_\alpha : \alpha \in \Lambda\}$.*

Proof Assume $x \in B \cap (\bigcup\{A_\alpha : \alpha \in \Lambda\})$. Then $x \in B$ and $x \in \bigcup\{A_\alpha : \alpha \in \Lambda\}$. Therefore there exists $\beta \in \Lambda$ such that $x \in A_\beta$. Hence $x \in B \cap A_\beta$. It follows that $x \in \bigcup\{B \cap A_\alpha : \alpha \in \Lambda\}$. This proves that $B \cap (\bigcup\{A_\alpha : \alpha \in \Lambda\}) \subseteq \bigcup\{B \cap A_\alpha : \alpha \in \Lambda\}$.

The proof that $\bigcup\{B \cap A_\alpha : \alpha \in \Lambda\} \subseteq B \cap (\bigcup\{A_\alpha : \alpha \in \Lambda\})$ is left as an exercise. ■

THEOREM 1.3.14 *Let $\{A_\alpha : \alpha \in \Lambda\}$ be an indexed collection of sets and let B be a set. Then $B \cup (\bigcap\{A_\alpha : \alpha \in \Lambda\}) = \bigcap\{B \cup A_\alpha : \alpha \in \Lambda\}$.*

Proof Assume $x \in \bigcap\{B \cup A_\alpha : \alpha \in \Lambda\}$. Then $x \in B \cup A_\alpha$ for every $\alpha \in \Lambda$. Now either $x \in B$ or $x \notin B$. If $x \in B$, then obviously $x \in B \cup (\bigcap\{A_\alpha : \alpha \in \Lambda\})$. Assume $x \notin B$. Since $x \in B \cup A_\alpha$ for every $\alpha \in \Lambda$, it follows that $x \in A_\alpha$ for every $\alpha \in \Lambda$. Therefore $x \in \bigcap\{A_\alpha : \alpha \in \Lambda\}$ which implies that $x \in B \cup (\bigcap\{A_\alpha : \alpha \in \Lambda\})$. Thus in either case we have that $x \in B \cup (\bigcap\{A_\alpha : \alpha \in \Lambda\})$. This proves that $\bigcap\{B \cup A_\alpha : \alpha \in \Lambda\} \subseteq B \cup (\bigcap\{A_\alpha : \alpha \in \Lambda\})$.

The proof that $B \cup (\bigcap\{A_\alpha : \alpha \in \Lambda\}) \subseteq \bigcap\{B \cup A_\alpha : \alpha \in \Lambda\}$ is left as an exercise. ■

Exercises 1.3

1. For each $i \in \{1, 2, 3, 4, 5\}$ let $A_i = [0, 1 + i)$. Find $A_1 \cap A_2 \cap A_3 \cap A_4 \cap A_5$ and $A_1 \cup A_2 \cup A_3 \cup A_4 \cup A_5$.
2. Let $\Lambda = \mathbb{R}^+$ and for each $\alpha \in \Lambda$ let $A_\alpha = [-2, \alpha)$. Find $\bigcup\{A_\alpha : \alpha \in \Lambda\}$ and $\bigcap\{A_\alpha : \alpha \in \Lambda\}$.
3. Let $\Lambda = \mathbb{Z}^+$ and for each $\alpha \in \Lambda$ let $B_\alpha = (\alpha - 1, \alpha + 1)$. Find $\bigcup\{B_\alpha : \alpha \in \Lambda\}$ and $\bigcap\{B_\alpha : \alpha \in \Lambda\}$.
4. Let $\Lambda = [0, 1]$ and for each $\alpha \in \Lambda$ let $C_\alpha = [-2, 1 + \alpha)$. Find $\bigcup\{C_\alpha : \alpha \in \Lambda\}$ and $\bigcap\{C_\alpha : \alpha \in \Lambda\}$.
5. Let $\Lambda = [0, 1)$ and for each $\alpha \in \Lambda$ let $D_\alpha = \{1, \alpha\}$. Find $\bigcup\{D_\alpha : \alpha \in \Lambda\}$ and $\bigcap\{D_\alpha : \alpha \in \Lambda\}$.
6. Let $A_1 = \{-1, 2\}$, $A_2 = \{-5\}$, and $A_3 = \{0, 2, 3\}$. List the elements of $A_1 \times A_2 \times A_3$.
7. Prove Theorem 1.3.12.
8. Complete the proof of Theorem 1.3.13.
9. Complete the proof of Theorem 1.3.14.
10. Let $\{A_\alpha : \alpha \in \Lambda\}$ be an indexed collection of sets. Prove the following statements:

    (a) for each $\beta \in \Lambda$, $A_\beta \subseteq \bigcup\{A_\alpha : \alpha \in \Lambda\}$.
    (b) for each $\beta \in \Lambda$, $\bigcap\{A_\alpha : \alpha \in \Lambda\} \subseteq A_\beta$.

11. Let $\{A_\alpha : \alpha \in \Lambda\}$ be an indexed collection of sets and let Δ be a nonempty subset of Λ. Prove the following statements:

(a) $\bigcup\{A_\alpha : \alpha \in \Delta\} \subseteq \bigcup\{A_\alpha : \alpha \in \Lambda\}$.
(b) $\bigcap\{A_\alpha : \alpha \in \Lambda\} \subseteq \bigcap\{A_\alpha : \alpha \in \Delta\}$.

1.4 *Functions*

Recall that the concept of a function plays an essential role in calculus. In fact, this concept is essential to all of mathematics and plays an especially important role in topology. Topology can actually be characterized as the study of certain types of functions. We begin this section by reviewing some basic notation and terms concerning functions.

Let X and Y be sets. A *function* f from X to Y is a rule or correspondence that associates each element of X with exactly one element of Y. For each $x \in X$ the unique element of Y associated with x by the function f is denoted by $f(x)$. The element $f(x)$ is called the *image* of x, and x is called a *pre-image* of $f(x)$. A function is also called a *map* or *mapping*. We say that a function f maps x onto $f(x)$. If f is a function from X to Y, we write $f: X \to Y$ or $X \xrightarrow{f} Y$. The set X is called the *domain* of f, and Y is called the *codomain* of f. The set $\{y \in Y : y = f(x) \text{ for some } x \in X\}$ is called the range of f. (Some authors use the term "range" for the set that we have called the codomain.)

Example 1.4.1
Let $f: \mathbb{R} \to \mathbb{R}$ be the function given by $f(x) = x^2 + 1$. The domain and the codomain of f are both $\mathbb{R}$. The range of f is $[1, +\infty)$.

Example 1.4.2
Let $g: \mathbb{R} \to [-1, 1]$ be given by $g(x) = \sin(x)$. The domain of g is $\mathbb{R}$, and both the codomain and range of g are $[-1, 1]$.

Example 1.4.3
Let $h: \mathbb{R} \to [-1, 5]$ be the function given by

$$h(x) = \begin{cases} -1 & \text{if } x < -2 \\ x^2 & \text{if } -2 \leqq x \leqq 2 \\ 5 & \text{if } x > 2 \end{cases}$$

The domain of h is $\mathbb{R}$, and the codomain of h is $[-1, 5]$. The range of h is $\{-1\} \cup [0, 4] \cup \{5\}$.

There are two types of functions that are especially useful in the study of topology. These are called one-to-one functions (or injective functions) and onto functions (or surjective functions).

Definition 1.4.4
A function $f: X \to Y$ is said to be *one-to-one* (or *injective*) provided that if $f(x_1) = f(x_2)$ then $x_1 = x_2$ for all $x_1, x_2 \in X$.

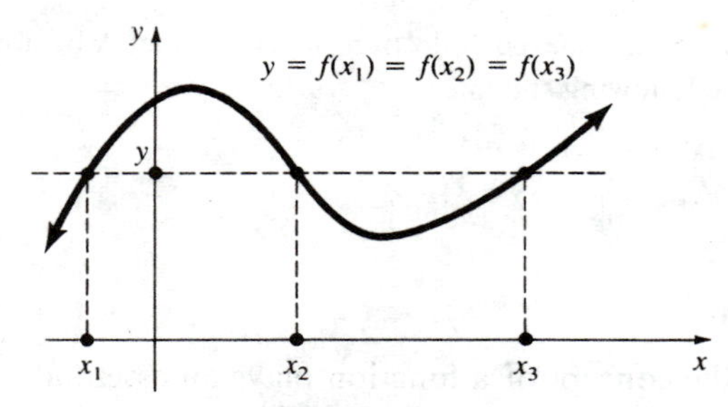

Figure 1.4.1
Graph of a function $f: \mathbb{R} \to \mathbb{R}$ that is not one-to-one.

In other words, a function is one-to-one if each element in the range is associated with only one element of the domain. For functions from $\mathbb{R}$ to $\mathbb{R}$ there is a geometric interpretation of the one-to-one property. A function $f: \mathbb{R} \to \mathbb{R}$ is one-to-one if any horizontal line intersects the graph of f in at most one point (see Figure 1.4.1).

Example 1.4.5
Let $f: \mathbb{R} \to \mathbb{R}$ be given by $f(x) = x^3 + 1$. In order to determine if f is one-to-one, assume that $f(x_1) = f(x_2)$, where $x_1, x_2 \in \mathbb{R}$. Then $x_1^3 + 1 = x_2^3 + 1$ and hence $x_1^3 = x_2^3$. Thus $x_1 = x_2$ and therefore f is one-to-one.

Example 1.4.6
Let $g: \mathbb{R} \to \mathbb{R}$ be the function defined by $g(x) = x^4 - x^2$. Since $g(-3) = g(3)$ and obviously $-3 \neq 3$, g is not one-to-one.

Example 1.4.7
Assume the function $h: \mathbb{R} \to \mathbb{R}$ is given by $h(x) = \sin(x)$. Note that $h(0) = h(\pi)$, and therefore h is not one-to-one. However, if $k: [-\pi/2, \pi/2] \to \mathbb{R}$ is given by $k(x) = \sin(x)$, then k is one-to-one.

Definition 1.4.8
A function $f: X \to Y$ is said to be *onto* (or *surjective*) provided that for each $y \in Y$ there exists at least one $x \in X$ such that $f(x) = y$.

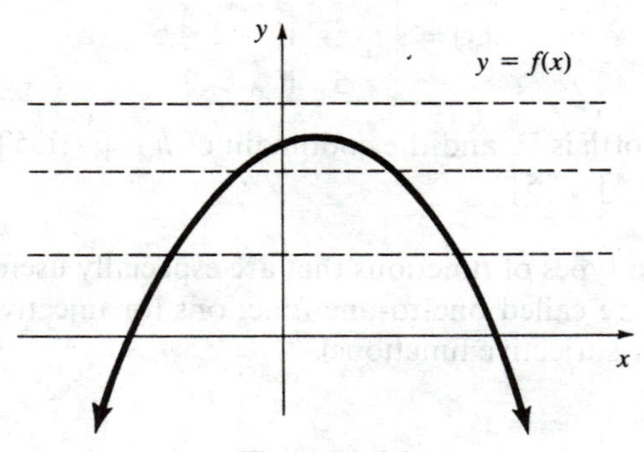

Figure 1.4.2
Graph of a function $f: \mathbb{R} \to \mathbb{R}$ that is not onto.

Thus a function is onto if the range is equal to the codomain. For functions from $\mathbb{R}$ to $\mathbb{R}$ there is a geometrical interpretation analogous to that for one-to-one functions. A function $f: \mathbb{R} \to \mathbb{R}$ is onto if every horizontal line intersects the graph of f in at least one point (see Figure 1.4.2).

Example 1.4.9

Assume the function $f: \mathbb{R} \to \mathbb{R}$ is defined by $f(x) = x^3 + 2$. To see that f is onto, let $y \in \mathbb{R}$. Then solving $y = x^3 + 2$ for x, we obtain $x = \sqrt[3]{y - 2}$. Thus $f(\sqrt[3]{y - 2}) = y$, and therefore f is onto.

Example 1.4.10

Let $g: \mathbb{R} \to \mathbb{R}$ be given by $g(x) = \sin(x)$. The function g is not onto because 2 is not in the range of g. That is, the equation $g(x) = 2$ has no solution. Note that the function $h: \mathbb{R} \to [-1, 1]$ defined by $h(x) = \sin(x)$ is onto. Whether or not a function is onto depends upon both the formula for the function and the codomain of the function.

Example 1.4.11

Assume $k: \mathbb{R} \to \mathbb{R}$ is given by

$$k(x) = \begin{cases} x + 2 & \text{if } x \leqq -1 \\ x^2 & \text{if } -1 < x < 1 \\ x + 1 & \text{if } x \geqq 1. \end{cases}$$

The graph of k is given in Figure 1.4.3. The function k is not onto because the interval $(1, 2)$ is disjoint from the range of k.

The next definition gives a useful way of combining functions.

Definition 1.4.12

Let $f: X \to Y$ and $g: Y \to Z$ be functions. The *composite* of f and g is the function $g \circ f: X \to Z$ given by $g \circ f(x) = g(f(x))$.

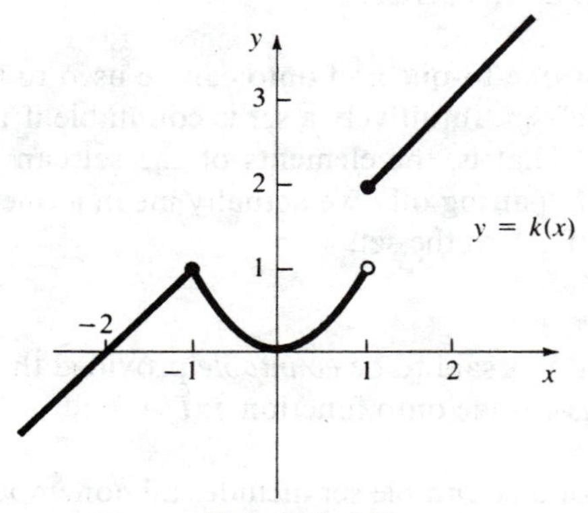

Figure 1.4.3

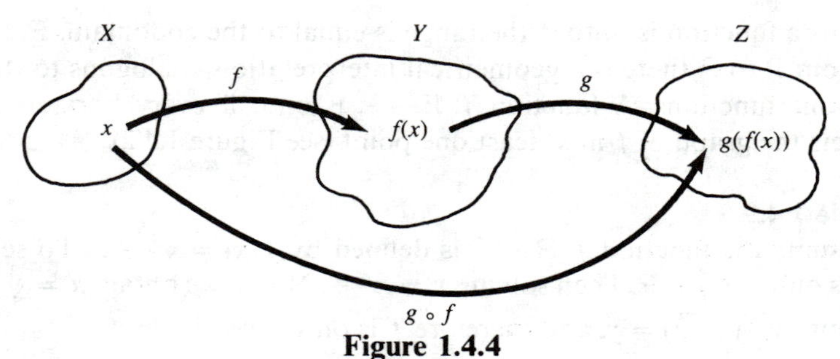

Figure 1.4.4

Note that in the above definition the codomain of f is the same as the domain of g. Actually the composite function $g \circ f$ is defined if the range of f is contained in the domain of g. An illustration of the concept of composition is given in Figure 1.4.4.

Example 1.4.13

Let $f: \mathbb{R} \to \mathbb{R}$ and $g: \mathbb{R} \to \mathbb{R}$ be given by $f(x) = 3x + 1$ and $g(x) = 2x^2 + x$. Then $g \circ f(x) = g(f(x)) = g(3x + 1) = 2(3x + 1)^2 + (3x + 1) = 18x^2 + 15x + 3$.

The following theorems state that the one-to-one and onto properties are preserved under composition.

THEOREM 1.4.14 *Let f: $X \to Y$ and g: $Y \to Z$ be functions. If both f and g are one-to-one, then $g \circ f$: $X \to Z$ is one-to-one.*

Proof Suppose that $g \circ f(x_1) = g \circ f(x_2)$, where $x_1, x_2 \in X$. Then $g(f(x_1)) = g(f(x_2))$. Since g is one-to-one, $f(x_1) = f(x_2)$. Similarly because f is one-to-one, $x_1 = x_2$. Therefore $g \circ f$ is one-to-one. ■

THEOREM 1.4.15 *Let f: $X \to Y$ and g: $Y \to Z$ be functions. If both f and g are onto, then $g \circ f$: $X \to Z$ is onto.*

The proof is left as an exercise.

The concepts of one-to-one and onto can be used to formally define the idea of a "countable" set. Intuitively a set is countable if the elements of the set can be counted. That is, the elements of the set can be paired off with positive integers. By "pairing off" we actually mean a one-to-one onto function from a subset of $\mathbb{Z}^+$ to the set.

Definition 1.4.16

A nonempty set X is said to be *countable* provided there exists a subset I of $\mathbb{Z}^+$ and a one-to-one onto function $f: I \to X$.

Our definition of a countable set includes all nonempty finite sets. It also includes some infinite sets. Obviously any subset of $\mathbb{Z}^+$ is countable. It can be shown that both $\mathbb{Z}$ and $\mathbb{Q}$ are countable. However $\mathbb{R}$ and the complex numbers are not countable. Any interval (consisting of more than one point) on the real number line is also not countable. The concept of a countable set is important in topology. However, because of the introductory nature of this text, the concept is used infrequently. For this course an intuitive notion of a countable set is sufficient.

It is often necessary to restrict the domain of a function. The next definition establishes the notation for this concept.

Definition 1.4.17

Let $f: X \to Y$ be a function and let $A \subseteq X$. The *restriction of f to A* is the function $f|_A: A \to Y$ given by $f|_A(x) = f(x)$ for $x \in A$.

In the above definition note that if $A \neq X$, then f and $f|_A$ are different functions because their domains are different.

Example 1.4.18

Let $f: \mathbb{R} \to \mathbb{R}$ be given by $f(x) = x^2 + 1$ and let $A = \mathbb{R}^+$. Then $f|_A: A \to \mathbb{R}$ is one-to-one, but f is not one-to-one.

The concept of an inverse is useful for functions. Intuitively the inverse of a function f is the function f "turned around." That is, the inverse of f maps elements in the range of f back to the corresponding elements in the domain of f. Of course, in order for the inverse of f to be a function, f must be one-to-one. In order to simplify the definition of the inverse of a function, we require that the function be one-to-one and onto.

Definition 1.4.19

Let $f: X \to Y$ be a one-to-one onto function. The *inverse* of f is the function $f^{-1}: Y \to X$ given by $f^{-1}(y) = x$ iff $f(x) = y$ (see Figure 1.4.5).

The next theorem relates the concept of inverse to that of composition.

Definition 1.4.20

The *identity function on a set X* is the function $\text{id}_X: X \to X$ given by $\text{id}_X(x) = x$ for each $x \in X$.

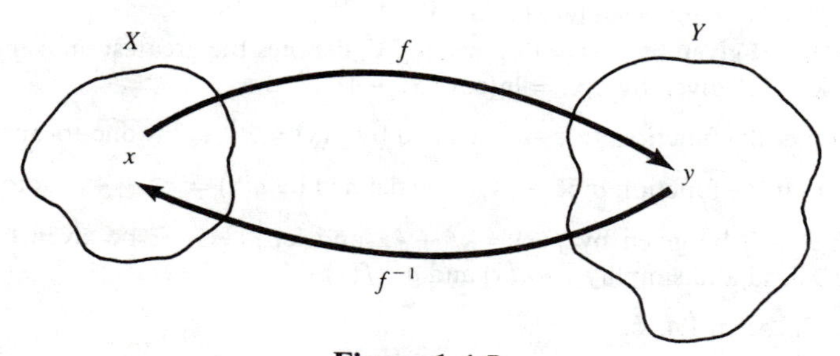

Figure 1.4.5

THEOREM 1.4.21 *Let $f\colon X \to Y$ and $g\colon Y \to X$ be one-to-one onto functions. Then $g = f^{-1}$ iff $f \circ g = \text{id}_Y$ and $g \circ f = \text{id}_X$.*

The proof is straightforward and is left as an exercise.

Example 1.4.22

Let $f\colon \mathbb{R} \to \mathbb{R}^+$ be given by $f(x) = e^x$ (where e is the irrational number that is the base for the natural logarithm). Then $f^{-1}(x) = \ln(x)$.

Example 1.4.23

Let $g\colon \mathbb{R} \to \mathbb{R}$ be given by $g(x) = x^3 - 5$. To find a formula for g^{-1}, we set $y = x^3 - 5$ and solve for x in terms of y. We obtain $x = \sqrt[3]{y+5}$. Therefore $g^{-1}\colon \mathbb{R} \to \mathbb{R}$ is given by $g^{-1}(y) = \sqrt[3]{y+5}$. Since we usually use x as the domain variable, we write $g^{-1}(x) = \sqrt[3]{x+5}$.

Exercises 1.4

1. Find the domain, codomain, and range of each of the following functions:

   (a) $f\colon \mathbb{R} \to \mathbb{R}$ given by $f(x) = 1 + \cos(x)$
   (b) $g\colon \mathbb{R} \to \mathbb{R}$ given by $g(x) = 4 - x^2$
   (c) $h\colon \mathbb{R} \to [3, +\infty)$ given by $h(x) = 4 + |x|$
   (d) $k\colon \mathbb{R} \to [-4, +\infty)$ given by

   $$k(x) = \begin{cases} -2 & \text{if } x < -1 \\ -x^2 & \text{if } -1 \leqq x \leqq 1 \\ x+2 & \text{if } x > 1 \end{cases}$$

2. Determine which of the following functions are one-to-one. If a particular function is not one-to-one, find two elements of the domain that have the same image.

   (a) $f\colon \mathbb{R} \to \mathbb{R}$ given by $f(x) = x^2 + 3x^4$
   (b) $g\colon [0, +\infty) \to \mathbb{R}$ given by $g(x) = x^2 + 3x^4$
   (c) $h\colon \mathbb{R} \to \mathbb{R}^+$ given by $h(x) = e^{x^3+8}$
   (d) $k\colon [0, 2\pi] \to \mathbb{R}$ given by $k(x) = \ln(\sin^2(x) + 1)$

3. Determine which of the following functions are onto. If a particular function is not onto, find an element of the codomain that is not in the range.

   (a) $f\colon \mathbb{R} \to \mathbb{R}$ given by $f(x) = 1 - x^2$
   (b) $g\colon \mathbb{R} \to [1, +\infty)$ given by $g(x) = x^2 + 1$
   (c) $h\colon \mathbb{R} \to \mathbb{R}$ given by $h(x) = [x]$, where $[x]$ denotes the greatest integer function
   (d) $k\colon \mathbb{R} \to \mathbb{R}^+$ given by $k(x) = \ln(\cos^2(x) + 4)$

4. Prove that the function $f\colon \mathbb{R} \to \mathbb{R}$ defined by $f(x) = x^5 + 7$ is one-to-one.

5. Prove that the function $g\colon \mathbb{R} \to [4, +\infty)$ defined by $g(x) = x^4 + 4$ is onto.

6. Let $f\colon \mathbb{R} \to \mathbb{R}$ be given by $f(x) = x^2 + 4x$ and let $g\colon \mathbb{R} \to \mathbb{R}$ be given by $g(x) = 8 - x^3$. Find and simplify $f \circ g(x)$ and $g \circ f(x)$.

7. Prove Theorem 1.4.15.

8. Prove that if $f: X \to Y$ and $g: Y \to Z$ are functions and $g \circ f$ is one-to-one, then f is one-to-one.
9. Prove that if $f: X \to Y$ and $g: Y \to Z$ are functions and $g \circ f$ is onto, then g is onto.
10. Let X and Y be sets with $A \subseteq X$. Prove that if a function $f: X \to Y$ is one-to-one, then $f|_A: A \to Y$ is one-to-one.
11. Find an example (different from any example in this section) of a function $f: X \to Y$ and $A \subseteq X$ such that $f|_A: A \to Y$ is one-to-one and f is not one-to-one.
12. Let X and Y be sets with $A \subseteq X$ and let $f: X \to Y$ be a function. Prove that if $f|_A: A \to Y$ is onto then $f: X \to Y$ is onto.
13. Find an example of a function $f: X \to Y$ and $A \subseteq X$ such that $f: X \to Y$ is onto and $f|_A: A \to Y$ is not onto.
14. Let $f: \mathbb{R} \to \mathbb{R}$ be given by $f(x) = 5x^3 + 9$. Find a formula for $f^{-1}: \mathbb{R} \to \mathbb{R}$.
15. Prove Theorem 1.4.21.

1.5 *Images and Inverse Images of Sets*

Up to this point we have considered images of single elements of the domain of a function. In this section images of subsets of the domain will be developed. We shall see that this is a very natural and useful extension of the notion of the image of an element. Inverse images of subsets of the codomain will also be investigated.

Definition 1.5.1

Let $f: X \to Y$ be a function and let $U \subseteq X$. The *image* of U is the set $f(U) = \{y \in Y : y = f(x) \text{ for some } x \in U\}$ (see Figure 1.5.1).

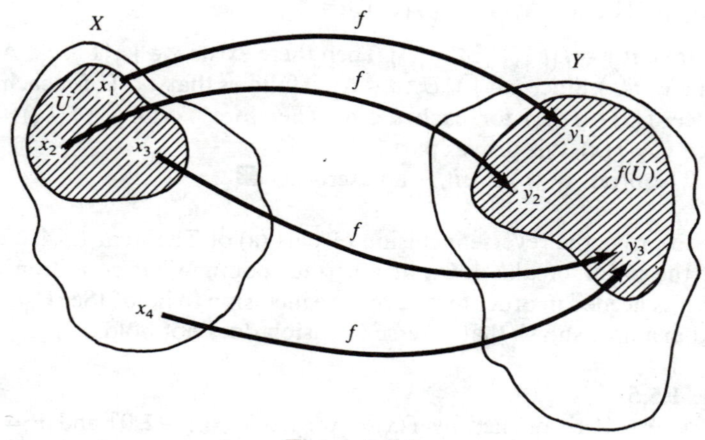

Figure 1.5.1

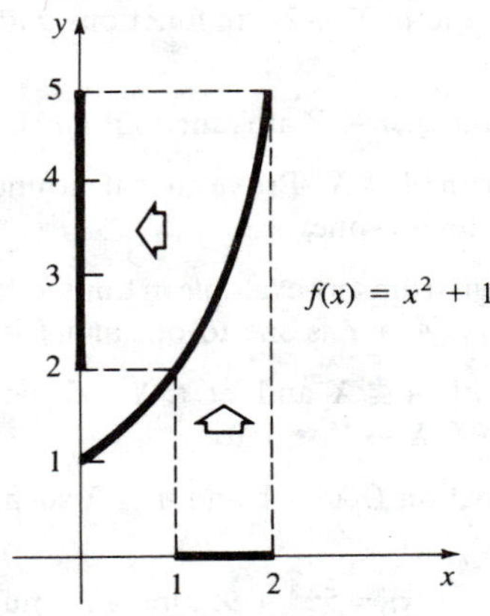

Figure 1.5.2

Example 1.5.2

Let $f: \mathbb{R} \to \mathbb{R}$ be given by $f(x) = x^2 + 1$. Then $f([1,2]) = [2,5]$ (see Figure 1.5.2). Similarly $f([-1,1]) = [1,2]$.

Example 1.5.3

Let $g: \mathbb{R} \to \mathbb{R}$ be given by $g(x) = 4 - x^2$. Then $g([0,2]) = [0,4]$, $g([-2,2]) = [0,4]$, and $g([2,+\infty)) = (-\infty, 0]$.

Next we consider how images of sets behave with respect to the operations of union and intersection of sets.

THEOREM 1.5.4 *Let $f: X \to Y$ be a function and let $\{U_\alpha : \alpha \in \Lambda\}$ be an indexed collection of subsets of X. Then*

(a) $f(\bigcap\{U_\alpha : \alpha \in \Lambda\}) \subseteq \bigcap\{f(U_\alpha) : \alpha \in \Lambda\}$

(b) $f(\bigcup\{U_\alpha : \alpha \in \Lambda\}) = \bigcup\{f(U_\alpha) : \alpha \in \Lambda\}$

Proof (a) Let $y \in f(\bigcap\{U_\alpha : \alpha \in \Lambda\})$. Then there exists $x \in \bigcap\{U_\alpha : \alpha \in \Lambda\}$ such that $y = f(x)$. Since $x \in \bigcap\{U_\alpha : \alpha \in \Lambda\}$, it follows that $x \in U_\alpha$ for each $\alpha \in \Lambda$. Hence $f(x) \in f(U_\alpha)$ for each $\alpha \in \Lambda$. Therefore $f(x) \in \bigcap\{f(U_\alpha) : \alpha \in \Lambda\}$. Since $y = f(x)$, this proves part (a).

The proof of (b) is left as an exercise. ■

Try to prove the reverse inclusion of part (a) of Theorem 1.5.4. You will see that the proof breaks down at a certain point. What condition on the function f is needed in order for the reverse inclusion to hold? (See Exercise 8.) The next example shows that reverse inclusion does not hold.

Example 1.5.5

Let $f: \mathbb{R} \to \mathbb{R}$ be defined by $f(x) = x^2$. Let $U = [-2,0]$ and $V = [0,2]$. Then $f(U \cap V) = f(\{0\}) = \{0\}$. However, $f(U) \cap f(V) = [0,4] \cap [0,4] = [0,4]$.

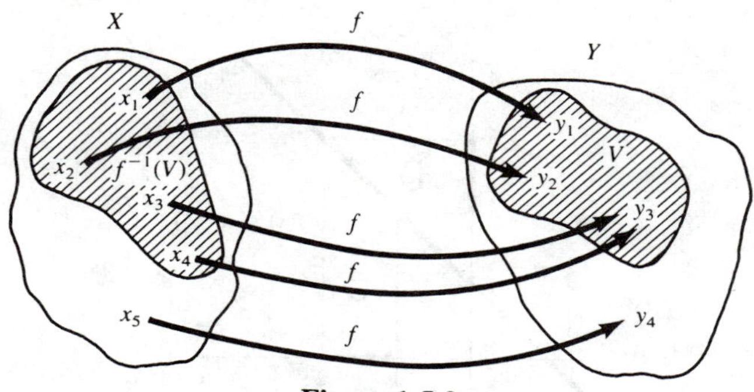

Figure 1.5.3

Example 1.5.6

Let $f\colon \mathbb{R} \to \mathbb{R}$ be given by $f(x) = 9 - x^2$. Let $\Lambda = \mathbb{R}^+$ and for each $\alpha \in \Lambda$ let $U_\alpha = [0, 1 + \alpha)$. Then $\bigcup\{f(U_\alpha) : \alpha \in \Lambda\} = f(\bigcup\{U_\alpha : \alpha \in \Lambda\}) = f([0, +\infty)) = (-\infty, 9]$.

Next we investigate inverse images of subsets of the codomain of a function. Recall that a function must be one-to-one in order for its inverse to be defined. We shall see that this is *not* the case for inverse images of subsets.

Definition 1.5.7

Let $f\colon X \to Y$ be a function and let $V \subseteq Y$. The *inverse image* of V is the set $f^{-1}(V) = \{x \in X : f(x) \in V\}$ (see Figure 1.5.3).

In other words, $f^{-1}(V)$ is the set of all x in X that are mapped to an element of V by the function f. The inverse image of a singleton set $\{y\}$ is usually denoted by $f^{-1}(y)$ rather than by $f^{-1}(\{y\})$. However, be very careful with this notation. In general $f^{-1}(y)$ denotes a *subset* of the domain, not necessarily a single element. In fact, if y is not in the range of f, then $f^{-1}(y)$ will be the empty set. The inverse image of any subset of the codomain is defined for any function. The function does not have to be either one-to-one or onto.

Example 1.5.8

Let $f\colon \mathbb{R} \to \mathbb{R}$ be defined by $f(x) = x + 1$. Then $f^{-1}([2, 3]) = [1, 2]$ (see Figure 1.5.4).

Example 1.5.9

Let $g\colon \mathbb{R} \to \mathbb{R}$ be given by $g(x) = 4 - x^2$. Then $g^{-1}([2, 3]) = [-\sqrt{2}, -1] \cup [1, \sqrt{2}]$ (see Figure 1.5.5).

In the following theorem we see that inverse images of sets are better behaved with respect to the operations of union and intersection than images of sets.

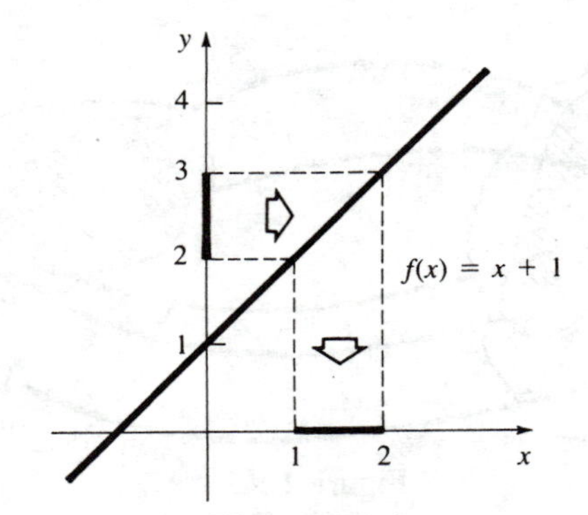

Figure 1.5.4

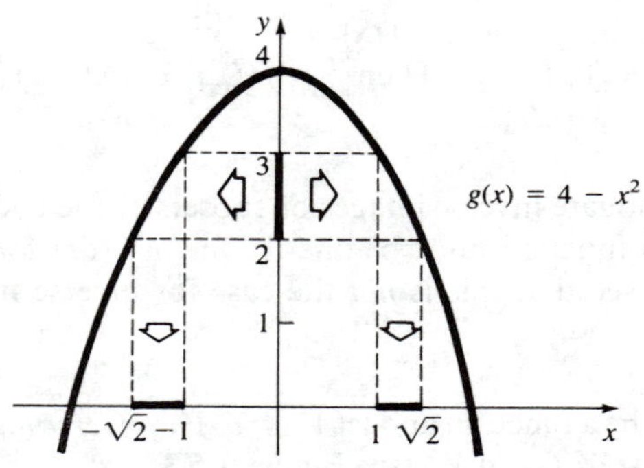

Figure 1.5.5

THEOREM 1.5.10 *Let $f: X \to Y$ be a function and let $\{V_\alpha : \alpha \in \Lambda\}$ be an indexed collection of subsets of Y. Then*

(a) $f^{-1}(\bigcap\{V_\alpha : \alpha \in \Lambda\}) = \bigcap\{f^{-1}(V_\alpha) : \alpha \in \Lambda\}$

(b) $f^{-1}(\bigcup\{V_\alpha : \alpha \in \Lambda\}) = \bigcup\{f^{-1}(V_\alpha) : \alpha \in \Lambda\}$.

Proof The proof of (a) is left as an exercise. To see that (b) holds, assume that $x \in X$. Then $x \in f^{-1}(\bigcup\{V_\alpha : \alpha \in \Lambda\})$ iff $f(x) \in \bigcup\{V_\alpha : \alpha \in \Lambda\}$ iff there exists $\beta \in \Lambda$ such that $f(x) \in V_\beta$ iff there exists $\beta \in \Lambda$ such that $x \in f^{-1}(V_\beta)$ iff $x \in \bigcup\{f^{-1}(V_\alpha) : \alpha \in \Lambda\}$. This completes the proof of part (b). ■

Example 1.5.11

Let $f: \mathbb{R} \to \mathbb{R}$ be defined by $f(x) = x^2 + 1$ and let $\Lambda = (0, 1]$. For each $\alpha \in \Lambda$ let $V_\alpha = [5 - \alpha, 5 + \alpha]$. Then

$$\bigcap\{f^{-1}(V_\alpha) : \alpha \in \Lambda\} = f^{-1}(\bigcap\{V_\alpha : \alpha \in \Lambda\}) = f^{-1}(5) = \{-2, 2\}$$

and

$$\bigcup\{f^{-1}(V_\alpha) : \alpha \in \Lambda\} = f^{-1}(\bigcup\{V_\alpha : \alpha \in \Lambda\}) = f^{-1}([4, 6]) = [-\sqrt{5}, -\sqrt{3}] \cup [\sqrt{3}, \sqrt{5}].$$

Next we see that the inverse image of the complement of a set is the complement of the inverse of the set.

THEOREM 1.5.12 *Let $f: X \to Y$ be a function and let $V \subseteq Y$. Then $f^{-1}(Y - V) = X - f^{-1}(V)$.*

Proof Suppose $x \in f^{-1}(Y - V)$. Then $f(x) \in Y - V$ and hence $f(x) \notin V$. This implies that $x \notin f^{-1}(V)$ and therefore $x \in X - f^{-1}(V)$. This proves that $f^{-1}(Y - V) \subseteq X - f^{-1}(V)$. The proof of the reverse inclusion is left as an exercise. ∎

It may seem that the image of the inverse image of a set should be the original set. However, this is not the case. Similarly, the inverse image of the image of a set is not necessarily the original set. The inclusions which do hold between these sets are given in the following theorem.

THEOREM 1.5.13 *Let $f: X \to Y$ be a function and let $U \subseteq X$ and $V \subseteq Y$. Then*

(a) $f(f^{-1}(V)) \subseteq V$

(b) $U \subseteq f^{-1}(f(U))$

Proof (a) Suppose $y \in f(f^{-1}(V))$. Then there exists $x \in f^{-1}(V)$ for which $y = f(x)$. Since $x \in f^{-1}(V)$, it follows that $f(x) \in V$ and therefore $y \in V$. This proves that $f(f^{-1}(V)) \subseteq V$.

The proof of (b) is left as an exercise. ∎

Try to prove the reverse inclusions of parts (a) and (b) of Theorem 1.5.13. You will have difficulty with these proofs. What additional assumptions are needed in order to complete the proofs? (See Exercises 11 and 12.) The following example shows that the reverse inclusions do not hold.

Example 1.5.14

Let $f: \mathbb{R} \to \mathbb{R}$ be defined by $f(x) = x^2$. Then $f(f^{-1}([-4, 4])) = f([-2, 2]) = [0, 4]$. Thus $f(f^{-1}([-4, 4])) \neq [-4, 4]$. Also

$$f^{-1}(f([0, 2])) = f^{-1}([0, 4]) = [-2, 2].$$

Therefore $f^{-1}(f([0, 2])) \neq [0, 2]$.

Exercises 1.5

1. Let $f: \mathbb{R} \to \mathbb{R}$ be given by $f(x) = x^2 - 1$. Find each of the following sets:

(a) $f([1, 2])$ **(b)** $f([0, 1])$
(c) $f([-1, 1])$ **(d)** $f([-2, -1] \cup [2, 3])$

2. Let $g: \mathbb{R} \to \mathbb{R}$ be defined by $g(x) = 9 - x^2$. Find each of the following sets:

(a) $g^{-1}([0, 9])$ **(b)** $g^{-1}([8, 10])$
(c) $g^{-1}([5, 8])$ **(d)** $g^{-1}([-16, -7] \cup [1, 5])$

3. Let $f: \mathbb{R} \to \mathbb{R}$ be given by $f(x) = 4 - x^2$. Find each of the following sets:
   (a) $f^{-1}(f([1,2]))$ **(b)** $f(f^{-1}([2,9]))$
4. Prove part (b) of Theorem 1.5.4.
5. Prove part (a) of Theorem 1.5.10.
6. Let $f: X \to Y$ be a function and let $U \subseteq X$ and $V \subseteq X$. Prove that $f(U) - f(V) \subseteq f(U - V)$.
7. Let $f: X \to Y$ be a function and $A \subseteq Y$ and $B \subseteq Y$. Prove that $f^{-1}(A) - f^{-1}(B) = f^{-1}(A - B)$.
8. Let $f: X \to Y$ be a function and let $\{U_\alpha : \alpha \in \Lambda\}$ be an indexed collection of subsets of X. Prove that if f is one-to-one, then $f(\bigcap\{U_\alpha : \alpha \in \Lambda\}) = \bigcap\{f(U_\alpha) : \alpha \in \Lambda\}$.
9. Complete the proof of Theorem 1.5.12.
10. Prove part (b) of Theorem 1.5.13.
11. Let $f: X \to Y$ be a function. Prove that f is onto iff $f(f^{-1}(V)) = V$ for every $V \subseteq Y$.
12. Let $f: X \to Y$ be a function. Prove that f is one-to-one iff $U = f^{-1}(f(U))$ for every $U \subseteq X$.

Review Exercises 1

Mark each of the following statements true or false. Briefly explain each true statement and find a counterexample for each false statement.

1. If A and B are sets and $A = B$, then $A - B = \varnothing$.
2. If A and B are sets and $A - B = \varnothing$, then $A = B$.
3. If both A and B are the empty set, then $A \times B = \varnothing$.
4. If A and B are sets and $A \times B = \varnothing$, then $A = \varnothing$ and $B = \varnothing$.
5. If A, B, and C are sets, then $A - (B - C) = (A - B) - C$.
6. If A and B are subsets of X and $A \cap B \neq \varnothing$, then $B \not\subseteq X - A$.
7. Let $\{A_\alpha : \alpha \in \Lambda\}$ be an indexed collection of sets. If $\bigcap\{A_\alpha : \alpha \in \Lambda\} = \varnothing$, then for any distinct α and β in Λ $A_\alpha \cap A_\beta = \varnothing$.
8. Let $\{A_\alpha : \alpha \in \Lambda\}$ be an indexed collection of sets. If for any distinct α and β in Λ $A_\alpha \cap A_\beta = \varnothing$, then $\bigcap\{A_\alpha : \alpha \in \Lambda\} = \varnothing$.
9. If $f: X \to Y$ is a function and $f(x_1) = f(x_2)$, then $x_1 = x_2$.
10. If $f: X \to Y$ is a one-to-one function and $f(x_1) = f(x_2)$, then $x_1 = x_2$.
11. If $f: X \to Y$ is a function and V is a nonempty subset of Y, then $f^{-1}(V)$ is a nonempty subset of X.
12. If $f: X \to Y$ is an onto function and V is a nonempty subset of Y, then $f^{-1}(V)$ is a nonempty subset of X.
13. The inverse of the inverse of a one-to-one onto function is the original function.
14. Let $f: X \to Y$ be a function and let A and B be subsets of Y. If $f^{-1}(A) = f^{-1}(B)$, then $A = B$.

15. If $f: X \to Y$ is a function, then $f(X) = Y$.

16. If $f: X \to Y$ is onto, then $f(X) = Y$.

17. Inverse images of sets are only defined for one-to-one functions.

18. If $f: X \to Y$ is a function, then $f^{-1}(Y) = X$.

19. If $f: X \to Y$ is a function and U and V are subsets of X, then $f(U \cap V) = f(U) \cap f(V)$.

20. If $f: X \to Y$ is a function and U and V are subsets of X, then $f(U \cap V) \subseteq f(U) \cap f(V)$.

2

Topological Spaces

2.1 *Open Subsets of the Real Numbers*

As stated in Chapter 1 our motivation for developing the concept of a topological space is the extension and generalization of the concept of continuity. In order to do this, continuity for functions between sets of real numbers will be defined in a way that can be generalized to an abstract setting. This definition will involve special subsets of the real numbers called open sets. These sets are related to open intervals.

Definition 2.1.1
A subset U of $\mathbb{R}$ is called an *open set* if $U = \varnothing$ or if for each $x \in U$ there is an open interval I such that $x \in I \subseteq U$.

The open subsets of $\mathbb{R}$ can also be characterized in terms of the absolute value function. A subset U of $\mathbb{R}$ is open iff for each $x \in U$ there is a positive number ε such that if $|x - y| < \varepsilon$, then $y \in U$.

Example 2.1.2
The set $(0, 1)$ is an open subset of $\mathbb{R}$.

Example 2.1.3
The set $U = [0, 1)$ is not an open set. There is no open interval I for which $0 \in I \subseteq U$.

Example 2.1.4
The set $V = (0, 1) \cup \{2\}$ is not an open set. There is no open interval I for which $2 \in I \subseteq V$.

The following theorem gives a slightly different characterization of open subsets of $\mathbb{R}$.

THEOREM 2.1.5 *A subset V of $\mathbb{R}$ is open* iff *V is equal to a union of open intervals.*

Proof ($\Rightarrow$) Assume V is an open subset of $\mathbb{R}$. If V is the empty set, then V is trivially the union of an empty collection of open intervals. If V is nonempty, then for each $x \in V$ there is an open interval I_x such that $x \in I_x \subseteq V$. It is easily seen that $V = \bigcup\{I_x : x \in V\}$.

($\Leftarrow$) Assume there is a collection of open intervals $\{I_\alpha : \alpha \in \Lambda\}$ such that $V = \bigcup\{I_\alpha : \alpha \in \Lambda\}$. Let $x \in V$. Then there is some $\beta \in \Lambda$ for which $x \in I_\beta$. Clearly $I_\beta \subseteq V$. Hence V is an open subset of $\mathbb{R}$. ■

The concept of an open subset of $\mathbb{R}$ can be used to define continuity.

Definition 2.1.6

A function $f: \mathbb{R} \to \mathbb{R}$ is said to be *continuous* if for each open subset V of $\mathbb{R}$, $f^{-1}(V)$ is an open subset of $\mathbb{R}$.

Example 2.1.7

Let $f(x) = \begin{cases} -2 & \text{if } x < 0 \\ 2 & \text{if } x \geq 0 \end{cases}$. Then $f^{-1}((0,3)) = [0,\infty)$. Note that $(0,3)$ is an open set and that $[0,\infty)$ is not an open set. Hence f is not a continuous function.

Next we show that Definition 2.1.6 is actually a restatement of the usual ε-δ characterization of continuity given in calculus.

THEOREM 2.1.8 *A function $f: \mathbb{R} \to \mathbb{R}$ is continuous (by Definition 2.1.6)* iff *for each $x_0 \in \mathbb{R}$, given $\varepsilon > 0$ there exists $\delta > 0$ such that $|x - x_0| < \delta \Rightarrow |f(x) - f(x_0)| < \varepsilon$.*

Proof ($\Rightarrow$) Assume $f: \mathbb{R} \to \mathbb{R}$ is continuous (by Definition 2.1.6). Let $x_0 \in \mathbb{R}$. Let $\varepsilon > 0$. Then the interval $(f(x_0) - \varepsilon, f(x_0) + \varepsilon)$ is an open subset of $\mathbb{R}$. By Definition 2.1.6, $f^{-1}((f(x_0) - \varepsilon, f(x_0) + \varepsilon))$ is an open subset of $\mathbb{R}$. Since $x_0 \in f^{-1}((f(x_0) - \varepsilon, f(x_0) + \varepsilon))$, there exists an open interval $I = (a,b)$ such that $x_0 \in I \subseteq f^{-1}((f(x_0) - \varepsilon, f(x_0) + \varepsilon))$. Now let $\delta = \min.\{x_0 - a, b - x_0\}$. Then

$$\begin{aligned} |x - x_0| < \delta &\Rightarrow x \in I \Rightarrow x \in f^{-1}((f(x_0) - \varepsilon, f(x_0) + \varepsilon)) \\ &\Rightarrow f(x) \in (f(x_0) - \varepsilon, f(x_0) + \varepsilon) \Rightarrow |f(x) - f(x_0)| < \varepsilon. \end{aligned}$$

($\Leftarrow$) Assume for each $x_0 \in \mathbb{R}$ that given $\varepsilon > 0$ there exists $\delta > 0$ for which $|x - x_0| < \delta \Rightarrow |f(x) - f(x_0)| < \varepsilon$. Let V be an open subset of $\mathbb{R}$. In order to show that $f^{-1}(V)$ is an open set, suppose that $x_0 \in f^{-1}(V)$. Then $f(x_0) \in V$. Since V is an open set, there exists an open interval

$I = (a, b)$ such that $f(x_0) \in I \subseteq V$. Let $\varepsilon = \min.\{f(x_0) - a, b - f\{x_0)\}$. Then $(f(x_0) - \varepsilon, f(x_0) + \varepsilon) \subseteq I$. By assumption there exists $\delta > 0$ such that $|x - x_0| < \delta \Rightarrow |f(x) - f(x_0)| < \varepsilon$. It is left as an exercise to show that $x_0 \in (x_0 - \delta, x_0 + \delta) \subseteq f^{-1}(V)$. Since x_0 is an arbitrary point of $f^{-1}(V)$ and $f^{-1}(V)$ contains an open interval about x_0, it follows that $f^{-1}(V)$ is an open set. ■

We shall study continuity in detail in the next chapter, but first some of the properties of open sets must be developed.

THEOREM 2.1.9 *The sets $\varnothing$ and $\mathbb{R}$ are both open sets.*

Proof The set $\varnothing$ is open by definition and $\mathbb{R}$ obviously satisfies Definition 2.1.1. ■

THEOREM 2.1.10 *Let $\{U_\alpha : \alpha \in \Lambda\}$ be a collection of open subsets of $\mathbb{R}$. Then $\bigcup\{U_\alpha : \alpha \in \Lambda\}$ is an open subset of $\mathbb{R}$.*

Proof Assume $x \in \bigcup\{U_\alpha : \alpha \in \Lambda\}$. Then for some $\beta \in \Lambda$, $x \in U_\beta$. Since U_β is an open set, there is an open interval I for which $x \in I \subseteq U_\beta$. Clearly $U_\beta \subseteq \bigcup\{U_\alpha : \alpha \in \Lambda\}$. Thus $x \in I \subseteq \bigcup\{U_\alpha : \alpha \in \Lambda\}$. Hence $\bigcup\{U_\alpha : \alpha \in \Lambda\}$ is an open set. ■

THEOREM 2.1.11 *Let $\{U_i : i = 1, 2, \ldots, n\}$ be a finite collection of open subsets of $\mathbb{R}$. Then $\bigcap\{U_i : i = 1, 2, \ldots, n\}$ is an open subset of $\mathbb{R}$.*

Proof Let $x \in \bigcap\{U_i : i = 1, 2, \ldots, n\}$. Then for each i, $x \in U_i$. Since each U_i is an open set, there is an open interval I_i for each i such that $x \in I_i \subseteq U_i$. Note that $x \in \bigcap\{I_i : i = 1, 2, \ldots, n\} \subseteq \bigcap\{U_i : i = 1, 2, \ldots, n\}$. By Exercise 7 $\bigcap\{I_i : i = 1, 2, \ldots, n\}$ is an open interval. Hence $\bigcap\{U_i : i = 1, 2, \ldots, n\}$ satisfies Definition 2.1.1 and is an open set. ■

The following example shows that the intersection of an infinite collection of open sets is not necessarily an open set.

Example 2.1.12
For each $n \in \mathbb{Z}^+$ let $U_n = (-1/n, 1/n)$. Then $\bigcap\{U_n : n \in \mathbb{Z}^+\} = \{0\}$ which is not an open set.

Exercises 2.1

1. Which of the following sets are open subsets of $\mathbb{R}$?

(a) $(0, 1)$ **(b)** $(0, 2]$ **(c)** $[2, 4]$
(d) $\{2\}$ **(e)** $(0, 1) \cup \{3\}$ **(f)** $\mathbb{Q}$
(g) $\mathbb{R} - \mathbb{Q}$ **(h)** $\{x \in \mathbb{R} : x^2 > 9\}$ **(i)** $(0, 2) \cup (3, 4)$
(j) $\mathbb{R} - \{0\}$

2. Show that any open interval is an open set.

3. Prove that the set $\mathbb{R} - \{1\}$ is an open set.
4. Prove that the set $\mathbb{R} - \{1, 2\}$ is an open set.
5. Prove that if F is any finite subset of $\mathbb{R}$, then $\mathbb{R} - F$ is an open set.
6. Show that the complement of the closed interval $[a, b]$ is an open set.
7. For $i = 1, 2, \ldots, n$ let $I_i = (a_i, b_i)$ be an open interval. Show that $\bigcap\{I_i : i = 1, 2, \ldots, n\}$ is either the empty set or an open interval.
8. Use Definition 2.1.6 to show that $f(x) = \begin{cases} -3 & \text{if } x < 1 \\ 3 & \text{if } x \geqq 1 \end{cases}$ is not a continuous function.
9. Use Definition 2.1.6 to show that $g(x) = \begin{cases} 1/x & \text{if } x \neq 0 \\ 0 & \text{if } x = 0 \end{cases}$ is not a continuous function.
10. Complete the proof of Theorem 2.1.8.

2.2 *Topological Spaces*

In this section the concept of an open set will be extended from the usual structure of the real number line to a more general setting. A collection of subsets of a set X will be defined to be the open subsets of X if the collection satisfies the same properties as the open subsets of $\mathbb{R}$. The characteristic properties of the open subsets of $\mathbb{R}$ were stated in Theorems 2.1.9, 2.1.10, and 2.1.11. This generalized notion of an open set will then be used to extend the concept of continuity.

Specifically, the collection of open subsets of a set X will be called a topology and the set together with this collection of subsets will be called a topological space. The concept of continuity will then be defined for functions between topological spaces. The following statement is the formal definition of a topological space.

Definition 2.2.1

Let X be a set. A collection $\mathscr{T}$ of subsets of X is called a *topology* if the following statements hold:

(a) $X \in \mathscr{T}$ and $\varnothing \in \mathscr{T}$.

(b) If $V_\alpha \in \mathscr{T}$ for each $\alpha \in \Lambda$, then $\bigcup\{V_\alpha : \alpha \in \Lambda\} \in \mathscr{T}$.

(c) If $V_i \in \mathscr{T}$ for $i = 1, 2, \ldots, n$, then $\bigcap\{V_i : i = 1, 2, \ldots, n\} \in \mathscr{T}$.

The sets in the collection $\mathscr{T}$ are called *open sets*. The pair $(X, \mathscr{T})$ is said to be a *topological space*.

The definition of a topology can be paraphrased as follows: A collection $\mathscr{T}$ of subsets of a set X is a topology for X if $\mathscr{T}$ contains $\varnothing$ and X, the union of any collection of sets in $\mathscr{T}$ is again in $\mathscr{T}$, and the intersection of any finite collection of sets in $\mathscr{T}$ is again in $\mathscr{T}$.

Note that the term "open set" in the context of a topological space $(X, \mathscr{T})$ simply means that the set is a member of the topology $\mathscr{T}$.

The open subsets of the real number line defined in Definition 2.1.1 will now be referred to as the *usual open subsets* of $\mathbb{R}$. This collection of subsets of $\mathbb{R}$ will be denoted by $\mathscr{U}$. We shall refer to $\mathscr{U}$ as the *usual topology* for $\mathbb{R}$. There are other topologies for $\mathbb{R}$. From this point on, the phrase "open subset of $\mathbb{R}$" must be used only in the context of a specific fixed topology for $\mathbb{R}$. A subset of $\mathbb{R}$ that is open with respect to one topology is not necessarily open with respect to another topology.

The notion of a topology is a very general concept. For any given set there may be many different topologies. If the particular topology in use is not clear from context, we shall say that a set is $\mathscr{T}$-open to mean that the set is a member of the topology $\mathscr{T}$. For example, the usual open subsets of $\mathbb{R}$ are the $\mathscr{U}$-open sets.

Example 2.2.2

The collection $\mathscr{H}$ of all subsets U of $\mathbb{R}$ such that either $U = \varnothing$ or for each $x \in U$ there is an interval of the form $[a, b)$ for which $x \in [a, b) \subseteq U$ is a topology for $\mathbb{R}$. We refer to this topology as the *half-open interval topology*.

The proof that $\mathscr{H}$ is a topology for $\mathbb{R}$ is similar to the proofs of Theorems 2.1.9, 2.1.10, and 2.1.11 and is left as an exercise.

Example 2.2.3

The collection $\mathscr{C} = \{(a, +\infty) : a \in \mathbb{R}\} \cup \{\mathbb{R}, \varnothing\}$ is a topology for $\mathbb{R}$. This topology is called the *open half-line topology*.

The next examples are topologies which can be defined for any set.

Example 2.2.4

Let X be any set. The collection $\mathscr{D}$ of all subsets of X is a topology for X. The collection $\mathscr{D}$ is called the *discrete topology*.

Example 2.2.5

Let X be any set. The collection $\mathscr{I} = \{X, \varnothing\}$ is a topology for X. This collection is called the *indiscrete topology* or *trivial topology*.

For a given set X the discrete topology and the indiscrete topology are the extreme topologies for X in the sense that $\mathscr{D}$ is the largest possible topology for X and $\mathscr{I}$ is the smallest possible topology for X, with respect to set inclusion.

The letters $\mathscr{U}$, $\mathscr{H}$, $\mathscr{C}$, $\mathscr{D}$, and $\mathscr{I}$ are used throughout this text to denote the usual topology, the half-open interval topology, the open half-line topology, the discrete topology, and the indiscrete topology, respectively.

Example 2.2.6

Let X be any nonempty set. Let $x \in X$. The collection of all subsets of X that are either empty or contain x is a topology for X. This topology is called the particular point topology.

Example 2.2.7

Let $X = \{a, b\}$ and let $\mathscr{T} = \{X, \varnothing, \{a\}\}$. The collection $\mathscr{T}$ is a special case of the particular point topology known as the Sierpinski topology.

Example 2.2.8

Let X be any set. The collection $\mathscr{T}$ of all subsets U of X such that either $U = \varnothing$ or $X - U$ is a finite set is a topology for X. This topology is called the finite complement topology or the cofinite topology.

Proof that the collection in Example 2.2.8 is a topology: Let X be any set and let $\mathscr{T} = \{U \subseteq X : U = \varnothing \text{ or } X - U \text{ is a finite set}\}$.

(a) By definition of $\mathscr{T}$, $\varnothing \in \mathscr{T}$ and since $X - X = \varnothing$, $X \in \mathscr{T}$.

(b) Let $U_\alpha \in \mathscr{T}$ for each $\alpha \in \Lambda$. If $U_\alpha = \varnothing$ for all $\alpha \in \Lambda$, then trivially $\bigcup\{U_\alpha : \alpha \in \Lambda\} = \varnothing \in \mathscr{T}$. Assume there is some $\beta \in \Lambda$ for which $U_\beta \neq \varnothing$. Then $X - \bigcup\{U_\alpha : \alpha \in \Lambda\} = \bigcap\{X - U_\alpha : \alpha \in \Lambda\} \subseteq X - U_\beta$. Since $U_\beta \neq \varnothing$, we have that $X - U_\beta$ is a finite set. Because $X - \bigcup\{U_\alpha : \alpha \in \Lambda\}$ is a subset of $X - U_\beta$, it must also be a finite set. Thus $\bigcup\{U_\alpha : \alpha \in \Lambda\} \in \mathscr{T}$.

(c) Let $U_1, U_2, \ldots, U_n$ be a finite collection of sets in $\mathscr{T}$. If for some i, $U_i = \varnothing$, then $\bigcap\{U_i : i = 1, 2, \ldots, n\} = \varnothing$ which is a member of $\mathscr{T}$. Assume that for all i $U_i \neq \varnothing$. Then for each i, $X - U_i$ is a finite set. Note that $X - \bigcap\{U_i : i = 1, 2, \ldots, n\} = \bigcup\{X - U_i : i = 1, 2, \ldots, n\}$ which is a union of a finite number of finite sets and hence finite. Thus $\bigcap\{U_i : i = 1, 2, \ldots, n\} \in \mathscr{T}$. Therefore $\mathscr{T}$ is a topology for X. ■

The following characterization will be useful in proving that a set is open.

THEOREM 2.2.9 *Let $(X, \mathscr{T})$ be a topological space. A subset A of X is open* iff *for each $x \in A$ there exists an open set U for which $x \in U \subseteq A$.*

Proof ($\Rightarrow$) Assume that A is an open set. Let $x \in A$. Then obviously if $U = A$, it follows that $x \in U \subseteq A$.

($\Leftarrow$) Assume that for each $x \in A$ there exists an open set U_x for which $x \in U_x \subseteq A$. Then $A = \bigcup\{U_x : x \in A\}$. Since for each $x \in A$, U_x is an open set and the union of any collection of open sets is again open, it follows that A is an open set. ■

Sometimes it is possible to compare different topologies for the same set.

Definition 2.2.10

Let X be a set and let $\mathscr{T}_1$ and $\mathscr{T}_2$ be topologies for X. If $\mathscr{T}_1 \subseteq \mathscr{T}_2$, then we say that $\mathscr{T}_2$ is *finer* than $\mathscr{T}_1$ or that $\mathscr{T}_1$ is *coarser* than $\mathscr{T}_2$.

Example 2.2.11

Let $\mathscr{T}$ be the finite complement topology for $\mathbb{R}$. Then $\mathscr{T} \subseteq \mathscr{U}$. In other words, $\mathscr{U}$ is finer than $\mathscr{T}$.

Example 2.2.12

Let X be any set and let $\mathscr{T}$ be any topology for X. Then $\mathscr{I} \subseteq \mathscr{T} \subseteq \mathscr{D}$, where $\mathscr{I}$ is the indiscrete topology for X and $\mathscr{D}$ is the discrete topology for X.

Since the concept of an open set has been extended from the usual structure of the real number line to a more general setting, continuity can now be extended to a general setting.

Definition 2.2.13

Let $(X, \mathscr{T})$ and $(Y, \mathscr{S})$ be topological spaces. A function $f: X \to Y$ is said to be *continuous* (or $\mathscr{T}-\mathscr{S}$ *continuous*) if for any $\mathscr{S}$-open subset V of Y, $f^{-1}(V)$ is a $\mathscr{T}$-open subset of X.

The continuity for functions from $\mathbb{R}$ to $\mathbb{R}$ given in Section 2.1 is actually $\mathscr{U}$-$\mathscr{U}$ continuity. Note that whether a function is continuous or not depends upon the topologies on the domain and codomain of the function. For example, a function $f: \mathbb{R} \to \mathbb{R}$ could be $\mathscr{H}$-$\mathscr{U}$ continuous but not $\mathscr{U}$-$\mathscr{U}$ continuous.

Example 2.2.14

Let $f: \mathbb{R} \to \mathbb{R}$ be given by

$$f(x) = \begin{cases} 3 & \text{if } x \geqq 0 \\ -3 & \text{if } x < 0 \end{cases}$$

Note that the set $(2, 4)$ is $\mathscr{U}$-open and that $f^{-1}((2, 4)) = [0, +\infty)$ which is not $\mathscr{U}$-open. Thus f is not $\mathscr{U}$-$\mathscr{U}$ continuous. Since the set $(2, 4)$ is also $\mathscr{H}$-open, it follows that f is not $\mathscr{U}$-$\mathscr{H}$ continuous. Also f is neither $\mathscr{C}$-$\mathscr{U}$ continuous nor $\mathscr{C}$-$\mathscr{H}$ continuous because $[0, +\infty)$ is not $\mathscr{C}$-open. Since $f^{-1}((1, +\infty)) = [0, +\infty)$, we have that f is not $\mathscr{C}$-$\mathscr{C}$ continuous. However, f is $\mathscr{H}$-$\mathscr{U}$ continuous, $\mathscr{H}$-$\mathscr{H}$ continuous, and $\mathscr{H}$-$\mathscr{C}$ continuous.

After the basic properties of topological spaces have been developed in this chapter, continuity will be investigated further in Chapter 3.

Exercises 2.2

1. Which of the following collections are topologies for $\mathbb{R}$? If a collection is not a topology for $\mathbb{R}$, explain why it is not.

(a) $\{\mathbb{R}, \varnothing, (-\infty, 0], (0, +\infty)\}$
(b) $\{\mathbb{R}, \varnothing, (-\infty, 1), (0, +\infty)\}$

(c) $\{\mathbb{R}, \varnothing, \{1\}\}$
(d) $\{U : U = \mathbb{R}$, or $U = \varnothing$ or $U = [a, +\infty)$ for some $a \in \mathbb{R}\}$
(e) $\{\mathbb{R}, \varnothing, (-\infty, 0], [0, +\infty)\}$
(f) $\{\mathbb{R}, \varnothing, (1, 3), (2, 4)\}$
(g) $\{\mathbb{R}, \varnothing, (1, 4), (2, 5), (1, 5)\}$

2. Let $X = \{a, b, c\}$. Which of the following collections are topologies for X? If a collection is not a topology, explain why it is not.

(a) $\{X, \{a\}, \{b\}, \{a, b\}\}$
(b) $\{X, \varnothing, \{a\}, \{b\}\}$
(c) $\{X, \varnothing, \{a\}\}$
(d) $\{X, \varnothing, \{a, b\}, \{b, c\}\}$
(e) $\{\{a\}, \{b\}, \{a, b\}\}$

3. Let X be a nonempty set. Let $x \in X$. Show that the collection

$$\mathcal{T} = \{U \subseteq X : U = \varnothing \text{ or } x \in U\}$$

is a topology for X. This is the particular point topology given in Example 2.2.6.

4. Let X be a nonempty set. Let $x \in X$. Show that $\mathcal{T} = \{U \subseteq X : U = X \text{ or } x \notin U\}$ is a topology for X. This collection is called the excluded point topology.

5. Show that the collection $\mathscr{H}$ in Example 2.2.2 is a topology for $\mathbb{R}$.

6. Prove that $\mathscr{H}$ is finer than $\mathscr{U}$. That is, show that every $\mathscr{U}$-open subset of $\mathbb{R}$ is also an $\mathscr{H}$-open set.

7. Show that $\mathscr{U}$ is finer than the finite complement topology for $\mathbb{R}$. (See Exercise 5 in Section 2.1)

8. Under what condition are the finite complement topology and the discrete topology the same topology for a fixed set?

9. Prove that the Sierpinski topology given in Example 2.2.7 satisfies Definition 2.2.1.

10. Show that the collection $\mathscr{C}$ given in Example 2.2.3 is a topology for $\mathbb{R}$.

11. Determine if the function $f: \mathbb{R} \to \mathbb{R}$ given by

$$f(x) = \begin{cases} 2 & \text{if } x \geqq 1 \\ -2 & \text{if } x < 1 \end{cases}$$

is (a) $\mathscr{U}$-$\mathscr{U}$ continuous
(b) $\mathscr{U}$-$\mathscr{H}$ continuous
(c) $\mathscr{U}$-$\mathscr{C}$ continuous
(d) $\mathscr{H}$-$\mathscr{U}$ continuous
(e) $\mathscr{H}$-$\mathscr{H}$ continuous
(f) $\mathscr{C}$-$\mathscr{H}$ continuous
(g) $\mathscr{C}$-$\mathscr{C}$ continuous.

12. Determine if the function $f: \mathbb{R} \to \mathbb{R}$ given by

$$f(x) = \begin{cases} 2 & \text{if } x > 1 \\ -2 & \text{if } x \leqq 1 \end{cases}$$

is (a) $\mathscr{U}$-$\mathscr{U}$ continuous
(b) $\mathscr{U}$-$\mathscr{H}$ continuous
(c) $\mathscr{U}$-$\mathscr{C}$ continuous
(d) $\mathscr{H}$-$\mathscr{U}$ continuous
(e) $\mathscr{H}$-$\mathscr{H}$ continuous

(f) $\mathscr{C}$-$\mathscr{H}$ continuous
(g) $\mathscr{C}$-$\mathscr{C}$ continuous.

2.3 *Closed Sets and Closure*

Definition 2.3.1
Let $(X, \mathscr{T})$ be a topological space. A subset U of X is said to be *closed* if $X - U$ is open.

Note that in any topological space $(X, \mathscr{T})$ the sets X and $\varnothing$ are both open and closed.

Example 2.3.2
Let $X = \{a, b, c\}$ and $\mathscr{T} = \{X, \varnothing, \{a\}, \{a, b\}\}$. The closed subsets are the complements of the sets in $\mathscr{T}$. That is, the closed sets are: $\varnothing$, X, $\{b, c\}$, and $\{c\}$.

Observe that the term "closed" is different from "not open." The set $\{a, c\}$ in the space in Example 2.3.2 is neither open nor closed.

Example 2.3.3
In the topological space $(\mathbb{R}, \mathscr{U})$ any closed interval $[a, b]$ is a closed set since $\mathbb{R} - [a, b] = (-\infty, a) \cup (b, +\infty)$ is an open set. A half-open interval of the form $[a, b)$ or $(a, b]$ is neither open nor closed.

Example 2.3.4
In the topological space $(\mathbb{R}, \mathscr{H})$ the half-open interval $[a, b)$ is both open and closed. The half-open interval $(a, b]$ is neither open nor closed.

Example 2.3.5
In the topological space $(\mathbb{R}, \mathscr{C})$ sets of the form $(-\infty, a]$ are closed because $\mathbb{R} - (-\infty, a] = (a, +\infty)$ which is open.

The terms "open" and "closed" are being used in several different contexts. In the context of the real numbers the terms "open interval," "closed interval," and "half-open interval" have their usual meaning. However, depending upon the topology on $\mathbb{R}$, any of these intervals may or may not be open sets or closed sets. Recall that in the context of a topological space, open sets are those sets in the topology and closed sets are those sets whose complements are in the topology.

Example 2.3.6
Let X be a nonempty set and let $x \in X$. Let $\mathscr{T} = \{U \subseteq X : U = \varnothing$ or $x \in U\}$. (The collection $\mathscr{T}$ is called the particular point topology.) The closed sets are the sets which do not contain x together with the set X.

THEOREM 2.3.7 *Let $(X, \mathscr{T})$ be a topological space and let $\{U_\alpha : \alpha \in \Lambda\}$ be a collection of closed subsets of X. Then $\bigcap\{U_\alpha : \alpha \in \Lambda\}$ is a closed set.*

Proof In order to show that $\bigcap\{U_\alpha : \alpha \in \Lambda\}$ is closed, we shall show its complement is open. Note that $X - \bigcap\{U_\alpha : \alpha \in \Lambda\} = \bigcup\{X - U_\alpha : \alpha \in \Lambda\}$. Since U_α is a closed set for each $\alpha \in \Lambda$, $X - U_\alpha$ is an open set for each $\alpha \in \Lambda$. Since an arbitrary union of open sets is an open set, $\bigcup\{X - U_\alpha : \alpha \in \Lambda\}$ is an open set. It follows that $\bigcap\{U_\alpha : \alpha \in \Lambda\}$ is a closed set. ∎

THEOREM 2.3.8 *Let $(X, \mathscr{T})$ be a topological space. Let*

$$\{U_i : i = 1, 2, \dots, n\}$$

be a finite collection of closed subsets of X. Then $\bigcup\{U_i : i = 1, 2, \dots, n\}$ is a closed set.

Proof In order to show that $\bigcup\{U_i : i = 1, 2, \dots, n\}$ is a closed set, we shall show that its complement is an open set. Observe that

$$X - \bigcup\{U_i : i = 1, 2, \dots, n\} = \bigcap\{X - U_i : i = 1, 2, \dots, n\}.$$

Since U_i is closed for each i, $X - U_i$ is open for each i. Then

$$\bigcap\{X - U_i : i = 1, 2, \dots, n\}$$

is the intersection of a finite number of open sets and hence is an open set. It follows that $\bigcup\{U_i : i = 1, 2, \dots, n\}$ is a closed set. ∎

The following example shows that the union of an infinite number of closed sets is not necessarily a closed set.

Example 2.3.9

For each $n \in \mathbb{Z}^+$ the set $U_n = [0, 1 - (1/n)]$ is a $\mathscr{U}$-closed set. However, $\bigcup\{U_n : n \in \mathbb{Z}^+\} = [0, 1)$ which is not a $\mathscr{U}$-closed set.

It is often useful to find the smallest closed set containing a given set.

Definition 2.3.10

Let $(X, \mathscr{T})$ be a topological space and let $A \subseteq X$. The *closure* of A is denoted by $\mathrm{Cl}(A)$ and is defined by $\mathrm{Cl}(A) = \bigcap\{U \subseteq X : U$ is a closed set and $A \subseteq U\}$.

The notation A^- is also commonly used for $\mathrm{Cl}(A)$. Note that by definition of $\mathrm{Cl}(A)$, we have $A \subseteq \mathrm{Cl}(A)$. Also $\mathrm{Cl}(A)$ is an intersection of closed sets and hence $\mathrm{Cl}(A)$ is a closed set. This makes $\mathrm{Cl}(A)$ the smallest closed set containing A.

Example 2.3.11

Let $X = \{a, b, c\}$ and $\mathscr{T} = \{X, \varnothing, \{a\}, \{b\}, \{a, b\}\}$. Let $A = \{b\}$. The closed sets are $\varnothing$, X, $\{b, c\}$, $\{a, c\}$, and $\{c\}$. The closed sets which contain A are X and $\{b, c\}$. Hence $\mathrm{Cl}(A) = X \cap \{b, c\} = \{b, c\}$.

Example 2.3.12
In the space $(\mathbb{R}, \mathscr{U})$ $\text{Cl}((0,1)) = [0,1]$.

Example 2.3.13
In the space $(\mathbb{R}, \mathscr{H})$ $\text{Cl}((0,1)) = [0,1)$. Note that $1 \notin \text{Cl}((0,1))$ because $[0,1)$ is a closed set containing $(0,1)$ but not the element 1.

Example 2.3.14
In the space $(\mathbb{R}, \mathscr{C})$ $\text{Cl}(\{1\}) = (-\infty, 1]$.

THEOREM 2.3.15 *Let $(X, \mathscr{T})$ be a topological space. If A is a closed subset of X, then $\text{Cl}(A) = A$.*

Proof From the definition of $\text{Cl}(A)$ it follows that $A \subseteq \text{Cl}(A)$.
Let $x \in \text{Cl}(A)$. Then every closed set containing A also contains x. Since A is a closed set containing A, it follows that $x \in A$. Therefore $\text{Cl}(A) \subseteq A$ and hence $\text{Cl}(A) = A$. ∎

COROLLARY 2.3.16 *In any topological space $(X, \mathscr{T})$, $\text{Cl}(\varnothing) = \varnothing$ and $\text{Cl}(X) = X$.*

THEOREM 2.3.17 *Let $(X, \mathscr{T})$ be a topological space with $A \subseteq X$ and $B \subseteq X$. Then*

(a) $\text{Cl}(A \cup B) = \text{Cl}(A) \cup \text{Cl}(B)$ *and*

(b) $\text{Cl}(A \cap B) \subseteq \text{Cl}(A) \cap \text{Cl}(B)$.

Proof (a) Let $x \in \text{Cl}(A \cup B)$. Suppose $x \notin \text{Cl}(A) \cup \text{Cl}(B)$. Therefore $x \notin \text{Cl}(A)$ and $x \notin \text{Cl}(B)$. Then there exists a closed set U containing A and a closed set V containing B such that $x \notin U$ and $x \notin V$. Thus $U \cup V$ is a closed set that contains $A \cup B$ but does not contain x. This implies that $x \notin \text{Cl}(A \cup B)$ which is a contradiction. Hence we must have that $x \in \text{Cl}(A) \cup \text{Cl}(B)$. Therefore $\text{Cl}(A \cup B) \subseteq \text{Cl}(A) \cup \text{Cl}(B)$.

Let $x \in \text{Cl}(A) \cup \text{Cl}(B)$. In order to show that $x \in \text{Cl}(A \cup B)$, we must show that every closed set containing $A \cup B$ also contains x. Let U be any closed set containing $A \cup B$. Now either $x \in \text{Cl}(A)$ or $x \in \text{Cl}(B)$. Assume $x \in \text{Cl}(A)$. Then since every closed set containing A also contains x, it follows that $x \in U$. Assume $x \in \text{Cl}(B)$. Then every closed set containing B also contains x. Thus $x \in U$. So in either case we have $x \in U$ and hence $x \in \text{Cl}(A \cup B)$. Therefore $\text{Cl}(A) \cup \text{Cl}(B) \subseteq \text{Cl}(A \cup B)$.

(b) Let $x \in \text{Cl}(A \cap B)$. In order to show that $x \in \text{Cl}(A) \cap \text{Cl}(B)$, we must show that $x \in \text{Cl}(A)$ and that $x \in \text{Cl}(B)$. That is, we must show that every closed set containing A also contains x and that every closed set containing B also contains x. Let U be any closed set containing A. Then obviously $A \cap B \subseteq U$. Since $x \in \text{Cl}(A \cap B)$, we must have $x \in U$. Therefore $x \in \text{Cl}(A)$. Similarly if V is any closed set containing B, then $A \cap B \subseteq V$ and it follows that $x \in V$. Hence $x \in \text{Cl}(B)$. Therefore

$x \in \mathrm{Cl}(A) \cap \mathrm{Cl}(B)$. Thus $\mathrm{Cl}(A \cap B) \subseteq \mathrm{Cl}(A) \cap \mathrm{Cl}(B)$. ■

The following example shows that in general $\mathrm{Cl}(A \cap B) \neq \mathrm{Cl}(A) \cap \mathrm{Cl}(B)$.

Example 2.3.18
In the topological space $(\mathbb{R}, \mathscr{U})$ let $A = (0, 1)$ and $B = (1, 2)$. Observe that $\mathrm{Cl}(A \cap B) = \mathrm{Cl}(\varnothing) = \varnothing$, but $\mathrm{Cl}(A) \cap \mathrm{Cl}(B)$ $[0, 1] \cap [1, 2] = \{1\}$.

Definition 2.3.19
Let $(X, \mathscr{T})$ be a topological space and let $A \subseteq X$. The set A is said to be *dense* in X provided that $\mathrm{Cl}(A) = X$.

Example 2.3.20
Let $X = \mathbb{R}$ and $\mathscr{T} = \{U \subseteq X : 1 \in U \text{ or } U = \varnothing\}$. The only closed set containing the set $\{1\}$ is X. Therefore $\mathrm{Cl}(\{1\}) = X$ and $\{1\}$ is a dense subset of $(X, \mathscr{T})$.

Exercises 2.3

1. Let $X = \{a, b, c\}$ and let $\mathscr{T} = \{X, \varnothing, \{b\}, \{a, b\}, \{b, c\}\}$.
   - **(a)** List the closed subsets of $(X, \mathscr{T})$.
   - **(b)** Find $\mathrm{Cl}(\{b\})$.
   - **(c)** Find $\mathrm{Cl}(\{a\})$.
   - **(d)** Find $\mathrm{Cl}(\{a, b\})$.
   - **(e)** Find a proper dense subset of X.
2. Show that $\{1, 2\}$ is a $\mathscr{U}$-closed subset of $\mathbb{R}$.
3. Show that any finite subset of $\mathbb{R}$ is $\mathscr{U}$-closed.
4. Show that in the space $(\mathbb{R}, \mathscr{U})$ $1 \notin \mathrm{Cl}((2, 3])$.
5. Prove that the set $[0, 1)$ is an $\mathscr{H}$-closed subset of $\mathbb{R}$.
6. Show that in the space $(\mathbb{R}, \mathscr{H})$ $1 \notin \mathrm{Cl}([0, 1))$.
7. Show that in the space $(\mathbb{R}, \mathscr{C})$ $0 \in \mathrm{Cl}([1, 2])$. Find $\mathrm{Cl}([1, 2])$ in the space $(\mathbb{R}, \mathscr{C})$.
8. Let $X = \mathbb{R}$ and let $\mathscr{T} = \{U \subseteq X : 1 \in U \text{ or } U = \varnothing\}$. Describe the closed subsets of X. Find $\mathrm{Cl}(\{1, 2\})$. Is $\{1, 2\}$ a dense subset of $(X, \mathscr{T})$? (The collection $\mathscr{T}$ is an example of the particular point topology.)
9. Let $X = \mathbb{R}$ and let $\mathscr{T} = \{U \subseteq X : 2 \notin U \text{ or } U = X\}$. Describe the closed subsets of X. Find $\mathrm{Cl}(\{3\})$. (The collection $\mathscr{T}$ is an example of the excluded point topology.)
10. Prove Corollary 2.3.16.
11. Show that if A is a subset of a topological space, then $\mathrm{Cl}(\mathrm{Cl}(A)) = \mathrm{Cl}(A)$.
12. Prove that in any topological space if U is an open set, then $\mathrm{Cl}(X - U) = X - U$.
13. Let U be a closed set and let V be an open set in a topological space. Show that $U - V$ is closed and that $V - U$ is open.

14. Let A and B be subsets of a topological space $(X, \mathscr{T})$. Show that $(X - \mathrm{Cl}(A)) \cup (X - \mathrm{Cl}(B)) \subseteq X - \mathrm{Cl}(A \cap B)$. Find an example that shows these sets are not in general equal.
15. Let A and B be subsets of a topological space $(X, \mathscr{T})$. Show that

$$X - \mathrm{Cl}(A \cup B) = (X - \mathrm{Cl}(A)) \cap (X - \mathrm{Cl}(B)).$$

2.4 *Limit Points, Interior, Exterior, Boundary, and More on Closure*

Definition 2.4.1

Let $(X, \mathscr{T})$ be a topological space with $A \subseteq X$. A point x in X is said to be a *limit point* of A provided that every open set containing x contains a point of A different from x. The set of all limit points of A is denoted by A'.

Limit points are also referred to as cluster points or accumulation points.

Example 2.4.2

Let $A = [0, 1)$ be a subset of the space $(\mathbb{R}, \mathscr{U})$. Then $A' = [0, 1]$. If A is a subset of the space $(\mathbb{R}, \mathscr{H})$, then $A' = [0, 1)$. Note that $1 \notin A'$ in the $\mathscr{H}$ topology since $[1, 2)$ is an $\mathscr{H}$-open set that contains 1 and is disjoint from A. In the $\mathscr{C}$ topology $A' = (-\infty, 1]$.

Example 2.4.3

Let $X = \mathbb{R}$ and $\mathscr{T} = \{U \subseteq X : 1 \in U \text{ or } U = \varnothing\}$. If $A = \{1\}$, then $A' = \mathbb{R} - \{1\}$.

In the following sequence of theorems the closure of a set is characterized in terms of the limit points of the set.

THEOREM 2.4.4 *Let $(X, \mathscr{T})$ be a topological space and let $A \subseteq X$. The set A is closed* iff $A' \subseteq A$.

Proof ($\Rightarrow$) Assume A is closed. Let $x \in A'$. Suppose $x \notin A$. Then $x \in X - A$. Since A is closed, $X - A$ is open. Therefore $X - A$ is an open set which contains x and is disjoint from A. This implies that $x \notin A'$ which is a contradiction. Thus $x \in A$ and therefore $A' \subseteq A$.

($\Leftarrow$) Assume $A' \subseteq A$. In order to show that A is closed, we shall show that $X - A$ is open. Let $y \in X - A$. Since $y \notin A$ and $A' \subseteq A$, it follows that $y \notin A'$. Then there exists an open set U containing y such that $U \cap A = \varnothing$. Thus $y \in U \subseteq X - A$. Hence $X - A$ is an open set. ■

Example 2.4.5

In the space $(\mathbb{R}, \mathscr{U})$ the set $[0, 1)$ is not closed because 1 is a limit point of $[0, 1)$ and is not a member of $[0, 1)$.

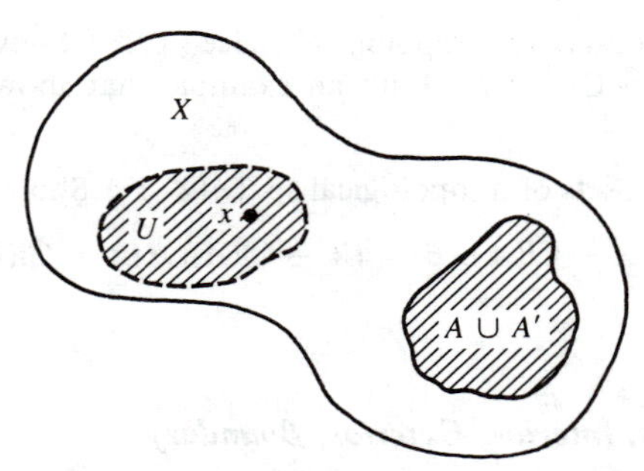

Figure 2.4.1

THEOREM 2.4.6 *Let $(X, \mathscr{T})$ be a topological space with $A \subseteq X$. The set $A \cup A'$ is closed.*

Proof In order to show that $A \cup A'$ is closed, we shall show that $X - A \cup A'$ is open. Let $x \in X - A \cup A'$. Then $x \notin A$ and $x \notin A'$. It follows that there exists an open set U such that $x \in U$ and $U \cap A = \varnothing$. Since U is open and disjoint from A, no point of U can be a limit point of A. Thus $U \cap A' = \varnothing$. Therefore $x \in U \subseteq X - A \cup A'$ and hence $X - A \cup A'$ is an open set (see Figure 2.4.1). ■

THEOREM 2.4.7 *Let $(X, \mathscr{T})$ be a topological space and let $A \subseteq X$. Then* $\text{Cl}(A) = A \cup A'$.

Proof Let $x \in A \cup A'$. Then $x \in A$ or $x \in A'$. If $x \in A$, then since $A \subseteq \text{Cl}(A)$, obviously $x \in \text{Cl}(A)$. Assume $x \in A'$. Recall that $\text{Cl}(A)$ is the intersection of all closed sets containing A. Let U be any closed set containing A. Since x is a limit point of A and $A \subseteq U$, we have that x is a limit point of U. Because U is closed, $x \in U$ by Theorem 2.4.4. It follows that $x \in \text{Cl}(A)$. Therefore $A \cup A' \subseteq \text{Cl}(A)$.

To see that $\text{Cl}(A) \subseteq A \cup A'$, note that by Theorem 2.4.6 $A \cup A'$ is a closed set. Obviously $A \cup A'$ contains the set A. Since $\text{Cl}(A)$ is the intersection of all closed sets containing A, we have that $\text{Cl}(A) \subseteq A \cup A'$. ■

Example 2.4.8
Let $X = \{a, b, c\}$ and $\mathscr{T} = \{X, \varnothing, \{a\}, \{a, b\}\}$. If $A = \{a\}$, then $A' = \{b, c\}$ and $\text{Cl}(A) = A \cup A' = X$.

The next theorem gives a useful characterization of the closure of a set.

THEOREM 2.4.9 *Let $(X, \mathscr{T})$ be a topological space and let $A \subseteq X$. Then* $x \in \text{Cl}(A)$ iff *for any open set U containing x, $U \cap A \neq \varnothing$.*

Proof ($\Rightarrow$) Let $x \in \text{Cl}(A)$. Then by Theorem 2.4.7 $x \in A$ or $x \in A'$. Let U be any open set containing x. If $x \in A$, obviously $U \cap A \neq \varnothing$. If $x \in A'$, then from the definition of a limit point it follows that $U \cap A \neq \varnothing$. Thus in either case $U \cap A \neq \varnothing$.

($\Leftarrow$) Assume that for each open set U containing x, $U \cap A \neq \emptyset$. If $x \in A$, then clearly $x \in \mathrm{Cl}(A)$. If $x \notin A$, then for any open set U containing x, $U \cap A$ contains a point of A different from x. Hence x is a limit point of A and $x \in \mathrm{Cl}(A)$. Therefore in either case $x \in \mathrm{Cl}(A)$. ■

Example 2.4.10
The closure of $\mathbb{Q}$ in the $\mathscr{U}$ topology is $\mathbb{R}$. That is, the rationals form a dense subset of $(\mathbb{R}, \mathscr{U})$. Note that every $\mathscr{U}$-open set containing an irrational number also contains a rational number and hence every irrational number is a limit point of the set of rational numbers.

Definition 2.4.11
Let $(X, \mathscr{T})$ be a topological space and let $A \subseteq X$. The *interior* of A, denoted by $\mathrm{Int}(A)$, is the set of all points $x \in X$ for which there exists an open set U such that $x \in U \subseteq A$.

Example 2.4.12
Let $X = \{a, b, c\}$ and $\mathscr{T} = \{X, \emptyset, \{a\}, \{a, b\}\}$. If $A = \{a, c\}$, then $\mathrm{Int}(A) = \{a\}$. If $B = \{b\}$, then $\mathrm{Int}(B) = \emptyset$.

Example 2.4.13
Let $A = [0, 1)$. In the $\mathscr{U}$ topology $\mathrm{Int}(A) = (0, 1)$. However, in the $\mathscr{H}$ topology $\mathrm{Int}(A) = [0, 1)$. With respect to the $\mathscr{C}$ topology, $\mathrm{Int}(A) = \emptyset$.

The following theorems give the basic properties of the interior of a set. The proofs are left as exercises.

THEOREM 2.4.14 *Let $(X, \mathscr{T})$ be a topological space with $A \subseteq X$. Then* $\mathrm{Int}(A) = \bigcup\{U \subseteq A : U \text{ is an open set}\}$.

THEOREM 2.4.15 *Let $(X, \mathscr{T})$ be a topological space. If $A \subseteq X$, then* $\mathrm{Int}(A)$ *is an open set.*

The next theorem implies that $\mathrm{Int}(A)$ is the largest open subset of A.

THEOREM 2.4.16 *Let $(X, \mathscr{T})$ be a topological space and let $A \subseteq X$. Then A is open* iff $A = \mathrm{Int}(A)$.

Definition 2.4.17
Let $(X, \mathscr{T})$ be a topological space and let $A \subseteq X$. The *exterior* of A, denoted by $\mathrm{Ext}(A)$, is the set of all points $x \in X$ for which there exists an open set U such that $x \in U \subseteq X - A$.

The proof of the following theorem is obvious and is left as an exercise.

THEOREM 2.4.18 *Let $(X, \mathscr{T})$ be a topological space and let $A \subseteq X$. Then* $\mathrm{Ext}(A) = \mathrm{Int}(X - A)$.

Example 2.4.19

Let $A = [0, 1)$. In the $\mathscr{U}$ topology $\text{Ext}(A) = (-\infty, 0) \cup (1, +\infty)$. However, in the $\mathscr{H}$ topology $\text{Ext}(A) = (-\infty, 0) \cup [1, +\infty)$. With respect to the $\mathscr{C}$ topology, $\text{Ext}(A) = (1, +\infty)$.

In Example 2.4.19 the points 0 and 1 are not in $\mathscr{U}$-Int(A) or $\mathscr{U}$-Ext(A). These points are in the set we shall call the ($\mathscr{U}$) boundary of A.

Definition 2.4.20

Let $(X, \mathscr{T})$ be a topological space and let $A \subseteq X$. The *boundary* of A, denoted by Bd(A), is the set of all points $x \in X$ for which every open set containing x intersects both A and $X - A$.

Some authors refer to the boundary of a set A as the frontier of A and use the notation Fr(A) for the boundary of A.

The next theorem gives the relationships among the interior, exterior, and boundary of a fixed set. The proof is left as an exercise.

THEOREM 2.4.21 *Let $(X, \mathscr{T})$ be a topological space and let $A \subseteq X$. Then the sets* Int(A), Bd(A), *and* Ext(A) *are pairwise disjoint and* $X = \text{Int}(A) \cup \text{Bd}(A) \cup \text{Ext}(A)$.

Example 2.4.22

Let $A = [0, 1) \cup (1, 2)$ be a subset of $(\mathbb{R}, \mathscr{U})$. Then $\text{Int}(A) = (0, 1) \cup (1, 2)$, $\text{Bd}(A) = \{0, 1, 2\}$, and $\text{Ext}(A) = (-\infty, 0) \cup (2, +\infty)$.

Example 2.4.23

Let $X = \mathbb{R}$ and $\mathscr{T} = \{U \subseteq \mathbb{R} : 1 \in U \text{ or } U = \varnothing\}$. Let $A = \{3, 4\}$. Then $\text{Int}(A) = \varnothing$, $\text{Bd}(A) = A$, and $\text{Ext}(A) = \mathbb{R} - A$.

Exercises 2.4

1. Let $X = \{a, b, c\}$ and $\mathscr{T} = \{X, \varnothing, \{a\}, \{a, b\}\}$. Let $A = \{a, c\}$. Find each of the following sets:

   (a) A' **(b)** $A \cup A'$ **(c)** Cl(A)

2. Let $A = (0, 1) \cup (1, 2]$ be a subset of $(\mathbb{R}, \mathscr{U})$. Find each of the following sets:

   (a) A' **(b)** $A \cup A'$ **(c)** Cl(A)

3. Let $A = [0, 1) \cup (2, 3) \cup \{4\}$ be a subset of $(\mathbb{R}, \mathscr{U})$. Find each of the following sets:

   (a) Cl(A) **(b)** Int(A) **(c)** Bd(A) **(d)** Ext(A)

4. Let $A = [0, 1) \cup (2, 3]$ be a subset of $(\mathbb{R}, \mathscr{H})$. Find each of the following sets:

   (a) Cl(A) **(b)** Int(A) **(c)** Bd(A) **(d)** Ext(A)

5. Let $A = (-\infty, 1) \cup [2, +\infty)$ be a subset of $(\mathbb{R}, \mathscr{C})$. Find each of the following sets:

   (a) Cl(A) **(b)** Int(A) **(c)** Bd(A) **(d)** Ext(A)

6. Let $X = \{a, b, c\}$ and $\mathscr{T} = \{X, \varnothing, \{a, b\}, \{a, c\}, \{a\}\}$. Let $A = \{a, c\}$. Find each of the following sets:

   (a) $\mathrm{Cl}(A)$ (b) $\mathrm{Int}(A)$ (c) $\mathrm{Bd}(A)$ (d) $\mathrm{Ext}(A)$

7. Let $X = \mathbb{R}$ and $\mathscr{T} = \{U \subseteq X : 2 \in U \text{ or } U = \varnothing\}$. Let $A = \{1, 3\}$. Find each of the following sets:

   (a) $\mathrm{Cl}(A)$ (b) $\mathrm{Int}(A)$ (c) $\mathrm{Bd}(A)$ (d) $\mathrm{Ext}(A)$

8. Prove Theorem 2.4.14.
9. Prove Theorem 2.4.15.
10. Show that if A is a subset of a topological space, then $\mathrm{Int}(\mathrm{Int}(A)) = \mathrm{Int}(A)$.
11. Prove Theorem 2.4.16.
12. Prove Theorem 2.4.18.
13. Prove Theorem 2.4.21.
14. Let A be a subset of a topological space. Prove that $\mathrm{Cl}(A) = \mathrm{Int}(A) \cup \mathrm{Bd}(A)$.
15. Show that for a subset A of a topological space, $\mathrm{Bd}(A) = \mathrm{Cl}(A) - \mathrm{Int}(A)$.
16. Let $A = (0, 1) \cup (1, 2)$ be a subset of $(\mathbb{R}, \mathscr{U})$. Find $\mathrm{Int}(\mathrm{Cl}(A))$. Is it true in general that if A is open, then $\mathrm{Int}(\mathrm{Cl}(A)) = A$?
17. Let $A = [0, 1] \cup \{2\}$ be a subset of $(\mathbb{R}, \mathscr{U})$. Find $\mathrm{Cl}(\mathrm{Int}(A))$. Is it true in general that if A is closed, then $\mathrm{Cl}(\mathrm{Int}(A)) = A$?

2.5 *Basic Open Sets*

Recall that the usual topology for $\mathbb{R}$ is defined in terms of open intervals. Specifically, a subset of $\mathbb{R}$ is $\mathscr{U}$-open iff it is a union of open intervals. These open intervals can be considered to be the "basic" open sets in $\mathscr{U}$. In an abstract topological space $(X, \mathscr{T})$ there are collections of $\mathscr{T}$-open sets which have the same relationship to the topology $\mathscr{T}$ as the open intervals do to the usual topology, $\mathscr{U}$.

Definition 2.5.1

Let $(X, \mathscr{T})$ be a topological space. A collection $\mathscr{B}$ of open sets is said to be a *base* for the topology $\mathscr{T}$ provided that each open set is a union of sets in $\mathscr{B}$. The sets in $\mathscr{B}$ are called *basic open* sets.

Example 2.5.2

The collection of all open intervals is a base for the $\mathscr{U}$ topology on $\mathbb{R}$. The collection of all half-open intervals of the form $[a, b)$ is a base for the $\mathscr{H}$ topology on $\mathbb{R}$. The collection of all intervals of the form $(a, +\infty)$ is a base for the $\mathscr{C}$ topology on $\mathbb{R}$.

A topology may have more than one base. Trivially any topology is a base for itself. However, we usually want a base containing as few sets as possible. The main ideas of working with a base instead of the whole topology are that there are fewer sets to consider and that the basic open sets are sometimes easier to describe.

Example 2.5.3
Each of the following collections is a base for the usual topology on $\mathbb{R}$:

(a) $\mathscr{B}_1 = \{(x - (1/n), x + (1/n)) : x \in \mathbb{R}, n \in \mathbb{Z}^+\}$
(b) $\mathscr{B}_2 = \{(a, b) : a, b \in \mathbb{R} \text{ and } b - a < 1\}$
(c) $\mathscr{B}_3 = \{(a, b) : a, b \in \mathbb{Q}\}$.

Example 2.5.4
The collection $\mathscr{B} = \{\{x\} : x \in \mathbb{R}\}$ is a base for $\mathscr{D}$ on $\mathbb{R}$.

Example 2.5.5
Let $X = \mathbb{R}$ and $\mathscr{T} = \{U \subseteq X : 1 \in U \text{ or } U = \varnothing\}$. The collection $\mathscr{B} = (\{\{x, 1\} : x \in \mathbb{R} - \{1\}\}) \cup \{\{1\}\}$ is a base for $\mathscr{T}$.

The following theorem gives a useful characterization of the basic open sets in a topology.

THEOREM 2.5.6 *Let $(X, \mathscr{T})$ be a topological space. A collection $\mathscr{B}$ of open sets is a base for $\mathscr{T}$* iff *for each open set U and each $x \in U$ there is a set $B \in \mathscr{B}$ for which $x \in B \subseteq U$.*

Proof ($\Rightarrow$) Assume $\mathscr{B}$ is a base for $\mathscr{T}$. Let U be an open set. Then U is equal to a union of sets in $\mathscr{B}$. That is, for some index set Λ, $U = \bigcup\{B_\alpha \in \mathscr{B} : \alpha \in \Lambda\}$. Let $x \in U$. Then there exists $\beta \in \Lambda$ for which $x \in B_\beta$. Obviously $B_\beta \subseteq U$.

($\Leftarrow$) Assume that for each open set U and each $x \in U$ there exists a set $B \in \mathscr{B}$ such that $x \in B \subseteq U$. Let U be an open set with $x \in U$. Then there exists a set $B_x \in \mathscr{B}$ for which $x \in B_x \subseteq U$. Therefore $U = \bigcup\{B_x : x \in U\}$. That is, U is a union of sets in $\mathscr{B}$ and hence $\mathscr{B}$ is a base for $\mathscr{T}$. ■

Any given collection of subsets of a fixed set X may or may not be a base for some topology on X. The next theorem states conditions under which a given collection of sets must be a base for some topology.

THEOREM 2.5.7 *Let X be a set. A collection $\mathscr{B}$ of subsets of X is a base for a topology on X* iff

(a) $X = \bigcup\{B : B \in \mathscr{B}\}$ *and*

(b) *for any $B_1, B_2 \in \mathscr{B}$ and any $x \in B_1 \cap B_2$, there exists $B_3 \in \mathscr{B}$ for which $x \in B_3 \subseteq B_1 \cap B_2$.*

Proof ($\Leftarrow$) Let $\mathscr{B}$ be a collection of subsets of X satisfying (a) and (b). Let $\mathscr{T}$ be the collection of all possible unions of sets in $\mathscr{B}$. We shall show that $\mathscr{T}$ is a topology for X. By (a) $X \in \mathscr{T}$ and since $\varnothing$ equals the union of the empty collection of sets in $\mathscr{B}$, it follows that $\varnothing \in \mathscr{T}$. Let $\{U_\alpha : \alpha \in \Lambda\}$

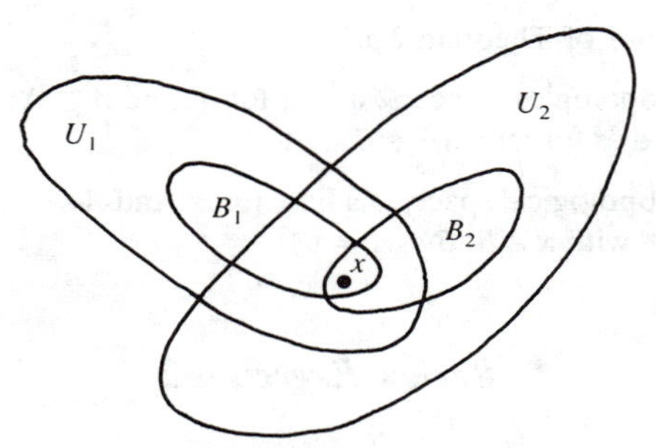

Figure 2.5.1

be a collection of sets in $\mathscr{T}$. Since each U_α is a union of sets in $\mathscr{B}$, clearly $\bigcup\{U_\alpha : \alpha \in \Lambda\}$ is a union of sets in $\mathscr{B}$. Thus by definition of $\mathscr{T}$, $\bigcup\{U_\alpha : \alpha \in \Lambda\} \in \mathscr{T}$. Let $\{U_i : i = 1, \ldots, n\}$ be a finite collection of sets in $\mathscr{T}$. We must show that $\bigcap\{U_i : i = 1, \ldots, n\} \in \mathscr{T}$. Let $x \in \bigcap\{U_i : i = 1, \ldots, n\}$. Then $x \in U_i$ for each i. Since each U_i is a union of sets in $\mathscr{B}$, it follows that for each i there exists $B_i \in \mathscr{B}$ such that $x \in B_i \subseteq U_i$ (see Figure 2.5.1). Thus $x \in \bigcap\{B_i : i = 1, \ldots, n\} \subseteq \bigcap\{U_i : i = 1, \ldots, n\}$. From (b) there exists a set $B_x \in \mathscr{B}$ for which $x \in B_x \subseteq \bigcap\{B_i : i = 1, \ldots, n\}$. Therefore for each $x \in \bigcap\{U_i : i = 1, \ldots, n\}$ we have a set $B_x \in \mathscr{B}$ for which $x \in B_x \subseteq \bigcap\{U_i : i = 1, \ldots, n\}$. It follows that $\bigcap(U_i : i = 1, \ldots, n\}$ is a union of sets in $\mathscr{B}$ and hence is a member of $\mathscr{T}$. Therefore $\mathscr{T}$ is a topology for X. It is clear from the definition of $\mathscr{T}$ that $\mathscr{B}$ is a base for $\mathscr{T}$.

The proof that a base for a topology satisfies (a) and (b) is left as an exercise. ■

Note that to prove a given collection $\mathscr{B}$ of $\mathscr{T}$-open sets is a base for a given topology $\mathscr{T}$, it is sufficient to show that each set in $\mathscr{T}$ is a union of sets in $\mathscr{B}$. However, to show that a given collection of sets $\mathscr{B}$ is a base for some topology, we must show that $\mathscr{B}$ satisfies (a) and (b) of Theorem 2.5.7.

Exercises 2.5

1. Explain why each of the following collections is not a base for a topology on $\mathbb{R}$.

   (a) $\{(n, n+1) : n \in \mathbb{Z}\}$ **(b)** $\{(x-1, x+1) : x \in \mathbb{R}\}$
   (c) $\{\{x, x+1\} : x \in \mathbb{R}\}$

2. Show that the collection $\mathscr{B}_3$ in Example 2.5.3 is a base for $\mathscr{U}$.

3. Show that $\mathscr{B} = \{(-x, x) : x \in \mathbb{R}^+\}$ is a base for a topology on $\mathbb{R}$. Is $\mathscr{B}$ a base for $\mathscr{U}$?

4. Show that $\mathscr{B} = \{\{x\} : x \in \mathbb{R}\}$ is a base for a topology on $\mathbb{R}$. Describe the topology.

5. Prove that any topology is a base for itself.

6. Show that the collection $\mathscr{B}$ in Example 2.5.5 is a base for $\mathscr{T} = \{U \subseteq \mathbb{R} : 1 \in U$ or $U = \varnothing\}$.

7. Show that $\mathscr{B} = \{(x, +\infty) : x \in \mathbb{Q}\}$ is a base for the $\mathscr{C}$ topology.

8. Complete the proof of Theorem 2.5.7.

9. Let $(X, \mathcal{T})$ be a topological space, $\mathcal{B}$ a base for $\mathcal{T}$, and $A \subseteq X$. Show that $x \in \text{Int}(A)$ iff there exists $B \in \mathcal{B}$ for which $x \in B \subseteq A$.

10. Let $(X, \mathcal{T})$ be a topological space, $\mathcal{B}$ a base for $\mathcal{T}$, and $A \subseteq X$. Show that $x \in \text{Cl}(A)$ iff for each $B \in \mathcal{B}$ with $x \in B$, $B \cap A \neq \varnothing$.

Review Exercises 2

Mark each of the following statements true or false. Briefly explain each true statement and find a counterexample for each false statement.

1. The empty set is a closed subset of $\mathbb{R}$ regardless of the topology on $\mathbb{R}$.

2. Any open interval is an open subset of $\mathbb{R}$ regardless of the topology on $\mathbb{R}$.

3. Any closed interval is a closed subset of $\mathbb{R}$ regardless of the topology on $\mathbb{R}$.

4. A half-open interval of the form $[a, b)$ is neither an open set nor a closed set regardless of the topology on $\mathbb{R}$.

5. If A is a subset of a topological space, then $A \subseteq \text{Cl}(A)$.

6. If A is a subset of a topological space, then $A' \subseteq A$.

7. For any closed subset A of a topological space, $A' \subseteq A$.

8. If A is a subset of a topological space, then $\text{Int}(A) \subseteq A$.

9. For any subset A of a topological space, $\text{Bd}(A) \subseteq A$.

10. If A is a subset of a topological space, then $\text{Bd}(A) \subseteq \text{Cl}(A)$.

11. If A is a closed subset of a topological space, then $\text{Bd}(A) \subseteq A$.

12. If A is a subset of a topological space, then $\text{Int}(A) \subseteq \text{Cl}(A)$.

13. The point 1 is a limit point of the set $[0, 1)$ regardless of the topology on $\mathbb{R}$.

14. The point 2 is not a limit point of the set $[0, 1)$ regardless of the topology on $\mathbb{R}$.

15. For any subset A of a topological space, $\text{Ext}(A) = X - A$.

16. For any closed subset A of a topological space, $\text{Ext}(A) = X - A$.

17. The collection $\mathcal{B} = \{\{x\} : x \in \mathbb{R}\}$ is a base for a topology on $\mathbb{R}$.

18. The collection $\mathcal{B} = \{\{x\} : x \in \mathbb{R}\}$ is a base for the usual topology on $\mathbb{R}$.

19. In a space $(X, \mathcal{T})$ any collection of open sets whose union equals X and that is closed under finite intersection is a base for $\mathcal{T}$.

20. There exists a topological space $(X, \mathcal{T})$ such that there is no base for $\mathcal{T}$.

21. There exists a topological space $(X, \mathcal{T})$ for which there is more than one base for $\mathcal{T}$.

3

Subspaces and Continuity

3.1 *Subspaces*

For any subset A of a topological space $(X, \mathscr{T})$, there is a natural topology for A induced by the topology $\mathscr{T}$.

THEOREM 3.1.1 *Let $(X, \mathscr{T})$ be a topological space and let $A \subseteq X$. The collection $\mathscr{T}_A = \{U \cap A : U \in \mathscr{T}\}$ is a topology for the set A.*

Proof Since $X, \varnothing \in \mathscr{T}$, it is clear that $A = X \cap A \in \mathscr{T}_A$ and $\varnothing = \varnothing \cap A \in \mathscr{T}_A$. Let $\{V_\alpha : \alpha \in \Lambda\}$ be a collection of sets in $\mathscr{T}_A$. For each $\alpha \in \Lambda$ there exists $U_\alpha \in \mathscr{T}$ for which $V_\alpha = U_\alpha \cap A$. Then $\bigcup\{V_\alpha : \alpha \in \Lambda\} = \bigcup\{U_\alpha \cap A : \alpha \in \Lambda\} = \bigcup\{U_\alpha : \alpha \in \Lambda\} \cap A$. Since $\bigcup\{U_\alpha : \alpha \in \Lambda\} \in \mathscr{T}$, it follows that $\bigcup\{V_\alpha : \alpha \in \Lambda\} \in \mathscr{T}_A$.

Let $V_1, V_2, \ldots, V_n$ be a finite collection of sets in $\mathscr{T}_A$. For each $i \in \{1, 2, \ldots, n\}$ there exists $U_i \in \mathscr{T}$ such that $V_i = U_i \cap A$. Observe that $\bigcap\{U_i \cap A : i = 1, 2, \ldots, n\} = \bigcap\{U_i : i = 1, 2, \ldots, n\} \cap A$. Because

$$\bigcap\{U_i : i = 1, 2, \ldots, n\} \in \mathscr{T},$$

we have that $\bigcap\{V_i : i = 1, 2, \ldots, n\} \in \mathscr{T}_A$.

Thus the collection $\mathscr{T}_A$ is a topology for A. ■

The topology $\mathscr{T}_A$ is called the *relative topology* for A and the space $(A, \mathscr{T}_A)$ is called a *subspace* of $(X, \mathscr{T})$. Also if $W \in \mathscr{T}_A$, we say that W is open relative to A or that W is open in A.

Example 3.1.2

Let $A = [0, 3)$ be a subspace of $(\mathbb{R}, \mathscr{U})$. Since $[0, 1) = A \cap (-1, 1)$ and $(-1, 1)$ is $\mathscr{U}$-open, it follows that $[0, 1)$ is $\mathscr{U}_A$-open. Note that $[0, 1)$ is not $\mathscr{U}$-open. The set $(1, 3)$ is both $\mathscr{U}_A$-open and $\mathscr{U}$-open. The set $(1, 2]$ is neither $\mathscr{U}_A$-open nor $\mathscr{U}$-open.

Example 3.1.3

Consider the subspace $A = (0,4]$ of $(\mathbb{R}, \mathscr{H})$. The set $[3,4]$ is $\mathscr{H}_A$-open since $[3,4] = [3,5) \cap A$ and $[3,5)$ is $\mathscr{H}$-open. Note that $[3,4]$ is not $\mathscr{H}$-open. The set $[2,4)$ is both $\mathscr{H}_A$-open and $\mathscr{H}$-open.

Example 3.1.4

Let $X = \{a,b,c\}$ and $\mathscr{T} = \{X, \varnothing, \{a\}, \{a,b\}\}$. If $A = \{a,c\}$, then $\mathscr{T}_A = \{A, \varnothing, \{a\}\}$.

Example 3.1.5

Let $X = \mathbb{R}$ and $\mathscr{T} = \{U \subseteq X : 1 \in U \text{ or } U = \varnothing\}$. Let $A = [2,5]$. For each $x \in A$ the set $\{1,x\}$ is $\mathscr{T}$-open. Since $\{x\} = \{1,x\} \cap A$, it follows that $\{x\}$ is $\mathscr{T}_A$-open. Thus $\mathscr{T}_A$ is the discrete topology.

Example 3.1.6

Let X be any set. If X has the discrete topology, then the relative topology for any subset of X is the discrete topology. Similarly if X has the indiscrete topology, then the relative topology for any subset of X is the indiscrete topology.

The proofs of the following theorems are left as exercises.

THEOREM 3.1.7 *Let $(X, \mathscr{T})$ be a topological space with $A \subseteq X$ and $U \subseteq A$. The set U is $\mathscr{T}_A$-closed* iff *$U = W \cap A$ for some $\mathscr{T}$-closed set W.*

For a subset A of a space $(X, \mathscr{T})$, the notation $\mathrm{Cl}(U)$ where $U \subseteq A$ is ambiguous. The closure can be taken with respect to either $(X, \mathscr{T})$ or $(A, \mathscr{T}_A)$ and the results are not necessarily the same. In order to alleviate this problem, the notations $\mathrm{Cl}_X(U)$ and $\mathrm{Cl}_A(U)$ will be used to denote the closure of U in X and A, respectively. A similar notation will be used for the Int, Ext, and Bd operators.

THEOREM 3.1.8 *Let $(X, \mathscr{T})$ be a topological space with $A \subseteq X$ and $U \subseteq A$. Then $\mathrm{Cl}_A(U) = A \cap \mathrm{Cl}_X(U)$.*

THEOREM 3.1.9 *Let $(X, \mathscr{T})$ be a topological space with $A \subseteq X$ and $U \subseteq A$. Then $A \cap \mathrm{Int}_X(U) \subseteq \mathrm{Int}_A(U)$.*

The next example shows that we do not have equality in Theorem 3.1.9.

Example 3.1.10

Let $A = [0,1]$ be a subspace of $(\mathbb{R}, \mathscr{U})$. Let $U = A$. Obviously $\mathrm{Int}_A(U) = \mathrm{Int}_A(A) = A = [0,1]$ and $A \cap \mathrm{Int}_{\mathbb{R}}(U) = [0,1] \cap (0,1) = (0,1)$.

THEOREM 3.1.11 *Let $(X, \mathscr{T})$ be a topological space with $A \subseteq X$ and $U \subseteq A$. Then $\mathrm{Bd}_A(U) \subseteq A \cap \mathrm{Bd}_X(U)$.*

It follows from the next example that $\mathrm{Bd}_A(U)$ is not necessarily equal to $A \cap \mathrm{Bd}_X(U)$.

Example 3.1.12

Let $A = [0,1]$ be a subspace of $(\mathbb{R}, \mathscr{U})$ and let $U = A$. Then $\mathrm{Bd}_A(U) = \mathrm{Bd}_A(A) = \varnothing$ and $A \cap \mathrm{Bd}_{\mathbb{R}}(U) = [0,1] \cap \{0,1\} = \{0,1\}$.

THEOREM 3.1.13 *If $\mathscr{B}$ is a base for a topological space $(X, \mathscr{T})$ and $A \subseteq X$, then the collection $\{B \cap A : B \in \mathscr{B}\}$ is a base for $(A, \mathscr{T}_A)$.*

THEOREM 3.1.14 *Let A be a subset of the space $(X, \mathscr{T})$. The set A is $\mathscr{T}$-open* iff *$\mathscr{T}_A \subseteq \mathscr{T}$.*

If we have the situation where $B \subseteq A \subseteq X$ and $(X, \mathscr{T})$ is a topological space, then there are two ways to construct a subspace topology for B. The following theorem states that both processes yield the same topology.

THEOREM 3.1.15 *Let $(X, \mathscr{T})$ be a topological space with $B \subseteq A \subseteq X$. Then $\mathscr{T}_B = (\mathscr{T}_A)_B$.*

Exercises 3.1

1. Let $A = [0,8)$ be a subspace of $(\mathbb{R}, \mathscr{U})$. Which of the following sets are $\mathscr{U}_A$-open? For each set V that is $\mathscr{U}_A$-open, find a $\mathscr{U}$-open set W for which $V = W \cap A$.

(a) $[0,2)$ **(b)** $(0,2]$ **(c)** $(2,3)$
(d) $[7,8)$ **(e)** $(7,8)$ **(f)** $[0,8)$
(g) $\varnothing$ **(h)** $\{4,5\}$ **(i)** $[2,6]$

2. Let $A = (0,8]$ be a subspace of $(\mathbb{R}, \mathscr{H})$. Which of the following sets are $\mathscr{H}_A$-open? For each set V that is $\mathscr{H}_A$-open, find an $\mathscr{H}$-open set W such that $V = W \cap A$.

(a) $(0,8)$ **(b)** $[6,8)$ **(c)** $[6,7)$
(d) $(0,1]$ **(e)** $[2,4]$ **(f)** $[2,4)$
(g) $(2,4)$ **(h)** $\varnothing$ **(i)** $(0,8]$

3. Let $A = (-3,0] \cup [1,3)$ be a subspace of $(\mathbb{R}, \mathscr{C})$. Which of the following sets are $\mathscr{C}_A$-open? For each set V that is $\mathscr{C}_A$-open, find a $\mathscr{C}$-open set W such that $V = W \cap A$.

(a) $(-2,0]$ **(b)** $(-2,0)$ **(c)** $(-2,0) \cup (1,2)$
(d) $(-2,0] \cup [1,3)$ **(e)** $(-2,0] \cup [2,3)$ **(f)** $(1,3)$
(g) $[1,3)$ **(h)** $(2,3)$ **(i)** $(-1,0] \cup [1,3)$

4. Let $X = \{a,b,c,d\}$ and $\mathscr{T} = \{X, \varnothing, \{a\}, \{b,c\}, \{a,b,c\}\}$. List the sets in each of the following topologies:

(a) $\mathscr{T}_{\{a,d\}}$ **(b)** $\mathscr{T}_{\{b,d\}}$ **(c)** $\mathscr{T}_{\{a,b,c\}}$ **(d)** $\mathscr{T}_{\{d\}}$

5. Let $X = \{a,b,c\}$ and $\mathscr{T} = \{X, \varnothing, \{a\}, \{b\}, \{a,b\}\}$. Let $A = \{a,c\}$. Find each of the following sets:

(a) $\mathrm{Int}_A(A)$ **(b)** $\mathrm{Int}_X(A)$ **(c)** $\mathrm{Int}_X(A) \cap A$

(d) $\mathrm{Bd}_A(A)$ (e) $\mathrm{Bd}_X(A)$ (f) $\mathrm{Bd}_X(A) \cap A$
(g) $\mathrm{Int}_A(\{a\})$ (h) $\mathrm{Int}_X(\{a\})$ (i) $\mathrm{Int}_A(\{c\})$

6. Let $X = \{a, b, c\}$ and $\mathscr{T} = \{X, \varnothing, \{a\}, \{a, b\}\}$. Let $A = \{a, c\}$. Find each of the following sets:

   (a) $\mathrm{Cl}_A(\{a\})$ (b) $\mathrm{Cl}_X(\{a\})$ (c) $\mathrm{Ext}_A(\{a\})$ (d) $\mathrm{Ext}_X(\{a\})$

7. Prove Theorem 3.1.7.

8. Let $(X, \mathscr{T})$ be a topological space and let $U \subseteq A \subseteq X$. Prove that if U is $\mathscr{T}$-open then U is $\mathscr{T}_A$-open.

9. Let $(X, \mathscr{T})$ be a topological space and let $V \subseteq A \subseteq X$. Prove that if V is $\mathscr{T}$-closed then V is $\mathscr{T}_A$-closed.

10. Prove Theorem 3.1.14.

11. Give and explain an example of a space $(X, \mathscr{T})$ and a subset A of X for which $\mathscr{T}_A \nsubseteq \mathscr{T}$.

12. Prove Theorem 3.1.13.

13. Let $\mathscr{B}$ be the base for $\mathscr{U}$ on $\mathbb{R}$ consisting of all open intervals. Let $A = [0, 5]$. Which of the following sets are in the base $\{B \cap A : B \in \mathscr{B}\}$ for $(A, \mathscr{T}_A)$?

    (a) $(0, 2)$ (b) $(0, 5]$ (c) $[0, 5]$
    (d) $[0, 5)$ (e) $(1, 3)$ (f) $[2, 4]$
    (g) $[2, 4) \cup (1, 3)$

14. Prove Theorem 3.1.8.

15. Prove Theorem 3.1.9.

16. Prove Theorem 3.1.11.

17. Prove Theorem 3.1.15.

3.2 *Continuity*

Continuity is one of the major reasons for developing the concept of a topological space. A topology is precisely the structure required on a set in order for continuity to be defined. The definition of continuity given in Chapter 2 (Definition 2.2.13) is restated here.

Definition 3.2.1

Let $(X, \mathscr{T})$ and $(Y, \mathscr{S})$ be topological spaces. A function $f\colon X \to Y$ is said to be *continuous* if for each $\mathscr{S}$-open subset V of Y, $f^{-1}(V)$ is a $\mathscr{T}$-open subset of X. If the topologies are not clear from context, the notation "$\mathscr{T}$-$\mathscr{S}$ continuity" will be used.

The concept of a "neighborhood" given in the following definition is a generalization of the notion of an open set and is useful in characterizing continuity.

Definition 3.2.2

Let $(X, \mathcal{T})$ be a topological space with $x \in X$. A *neighborhood* (or *$\mathcal{T}$-neighborhood*) of x is a subset N of X for which there is an open set U such that $x \in U \subseteq N$. (The abbreviation "ngbh." will sometimes be used for neighborhood.)

In other words, a neighborhood of x is a set that contains an open set which contains x. Obviously every open set containing x is a neighborhood of x. However, a neighborhood of x is not necessarily an open set.

Example 3.2.3

In $(\mathbb{R}, \mathcal{U})$ the set $[0, 2]$ is a neighborhood of 1, but $[0, 2]$ is clearly not open. The set $(-1, 1)$ is a neighborhood of 0 and also an open set. The set $[0, 1]$ is not a neighborhood of 0 since it does not contain an open set about 0.

Example 3.2.4

In $(\mathbb{R}, \mathcal{H})$ the set $[0, 1]$ is a neighborhood of 0 because $0 \in [0, 1) \subseteq [0, 1]$ and $[0, 1)$ is an open set. However $[0, 1]$ is not a neighborhood of 1.

Example 3.2.5

Let $X = \{a, b, c\}$ and $\mathcal{T} = \{X, \varnothing, \{a\}, \{a, b\}\}$. The set X is the only neighborhood of c. The set $\{a, c\}$ is a neighborhood of a but not a neighborhood of c. The sets $\{a, b\}$ and X are the only neighborhoods of b.

The following theorems characterize open sets, interiors of sets, and closures of sets in terms of neighborhoods. The proofs are left as exercises.

THEOREM 3.2.6 *A subset U of a topological space $(X, \mathcal{T})$ is open* iff *U is a neighborhood of each of its points.*

THEOREM 3.2.7 *Let $(X, \mathcal{T})$ be a topological space with $U \subseteq X$. Then* $\text{Cl}(U) = \{x \in X$*: if N is a ngbh. of x, then $N \cap U \neq \varnothing$*$\}$.

THEOREM 3.2.8 *Let $(X, \mathcal{T})$ be a topological space and let $U \subseteq X$. Then* $\text{Int}(U) = \{x \in X$*: U is a ngbh. of x*$\}$.

The following theorem gives several useful characterizations of continuity.

THEOREM 3.2.9 *Let $(X, \mathcal{T})$ and $(Y, \mathcal{S})$ be topological spaces and let $f: X \to Y$ be a function. The following statements are equivalent:*

(a) *The function f is continuous.*

(b) *For each closed subset V of Y, $f^{-1}(V)$ is a closed subset of X.*

(c) *If $\mathcal{B}$ is a base for $(Y, \mathcal{S})$, then for each $B \in \mathcal{B}$, $f^{-1}(B)$ is $\mathcal{T}$-open.*

(d) *If $x \in X$ and N is an $\mathscr{S}$-neighborhood of $f(x)$, then $f^{-1}(N)$ is a $\mathscr{T}$-neighborhood of x.*

(e) *If $x \in X$ and N is an $\mathscr{S}$-neighborhood of $f(x)$, then there is a $\mathscr{T}$-neighborhood M of x for which $f(M) \subseteq N$.*

(f) *If $U \subseteq X$, then $f(\mathrm{Cl}(U)) \subseteq \mathrm{Cl}(f(U))$.*

Proof We shall prove that (a) $\Rightarrow$ (c) $\Rightarrow$ (d) $\Rightarrow$ (e) $\Rightarrow$ (f) $\Rightarrow$ (b) $\Rightarrow$ (a).

(a) $\Rightarrow$ (c). Let $\mathscr{B}$ be a base for $(Y, \mathscr{S})$ and let $B \in \mathscr{B}$. Since B is $\mathscr{S}$-open, it follows from (a) that $f^{-1}(B)$ is $\mathscr{T}$-open.

(c) $\Rightarrow$ (d). Let $x \in X$ and let N be an $\mathscr{S}$-neighborhood of $f(x)$. Then there is an $\mathscr{S}$-open set V such that $f(x) \in V \subseteq N$. Let $\mathscr{B}$ be a base for $(Y, \mathscr{S})$. There is a set $B \in \mathscr{B}$ for which $f(x) \in B \subseteq V$. By (c), $f^{-1}(B)$ is $\mathscr{T}$-open. Since $x \in f^{-1}(B) \subseteq f^{-1}(V) \subseteq f^{-1}(N)$, it follows that $f^{-1}(N)$ is a $\mathscr{T}$-neighborhood of x.

(d) $\Rightarrow$ (e). Let $x \in X$ and let N be an $\mathscr{S}$-neighborhood of $f(x)$. By (d), $f^{-1}(N)$ is a $\mathscr{T}$-neighborhood of x. Thus there is a $\mathscr{T}$-open set U for which $x \in U \subseteq f^{-1}(N)$. Obviously U is a $\mathscr{T}$-neighborhood of x and $f(U) \subseteq f(f^{-1}(N)) \subseteq N$.

(e) $\Rightarrow$ (f). Let $y \in f(\mathrm{Cl}(U))$. Let V be any $\mathscr{S}$-open set containing y. In order to show that $y \in \mathrm{Cl}(f(U))$, we shall show that $V \cap f(U) \neq \varnothing$. Let $x \in \mathrm{Cl}(U)$ such that $y = f(x)$. Since V is a $\mathscr{S}$-neighborhood of $f(x)$, it follows from (e) that there is a $\mathscr{T}$-neighborhood M of x for which $f(M) \subseteq V$. Since $x \in \mathrm{Cl}(U)$, we have from Theorem 3.2.7 that $M \cap U \neq 0$. Let $z \in M \cap U$. Then $f(z) \in f(M \cap U) \subseteq f(M) \cap f(U) \subseteq V \cap f(U)$. Thus $V \cap f(U) \neq \varnothing$. Therefore $y \in \mathrm{Cl}(f(U))$ and hence $f(\mathrm{Cl}(U)) \subseteq \mathrm{Cl}(f(U))$.

(f) $\Rightarrow$ (b). Let V be a closed subset of Y. By (f), $f(\mathrm{Cl}(f^{-1}(V))) \subseteq \mathrm{Cl}(f(f^{-1}(V))) \subseteq \mathrm{Cl}(V) = V$. Therefore $\mathrm{Cl}(f^{-1}(V)) \subseteq f^{-1}(V)$. Thus $\mathrm{Cl}(f^{-1}(V)) = f^{-1}(V)$ and hence $f^{-1}(V)$ is a closed subset of X.

(b) $\Rightarrow$ (a). Let W be an open subset of Y. Then $Y - W$ is a closed set. By (b), $f^{-1}(Y - W)$ is a closed subset of X. Since $f^{-1}(Y - W) = X - f^{-1}(W)$, it follows that $f^{-1}(W)$ is an open subset of X. Thus f is continuous. ■

Example 3.2.10

Let $f\colon \mathbb{R} \to \mathbb{R}$ be given by

$$f(x) = \begin{cases} x + 1 & \text{if } x \geqq 1 \\ -1 & \text{if } x < 1 \end{cases}$$

Since the set $(1, 3)$ is $\mathscr{U}$-open and $f^{-1}((1, 3)) = [1, 2)$ (see Figure 3.2.1) which is not $\mathscr{U}$-open, f is not $\mathscr{U}$-$\mathscr{U}$ continuous. The set $(1, 3)$ is also $\mathscr{H}$-open, which implies that f is not $\mathscr{U}$-$\mathscr{H}$ continuous. Since $[1, 2)$ is not $\mathscr{C}$-open, f is neither $\mathscr{C}$-$\mathscr{U}$ continuous nor $\mathscr{C}$-$\mathscr{H}$ continuous. The function f is not $\mathscr{C}$-$\mathscr{C}$ continuous because $f^{-1}((0, +\infty)) = [1, +\infty)$ which is not $\mathscr{C}$-open. However, f is both $\mathscr{H}$-$\mathscr{U}$ continuous and $\mathscr{H}$-$\mathscr{H}$ continuous.

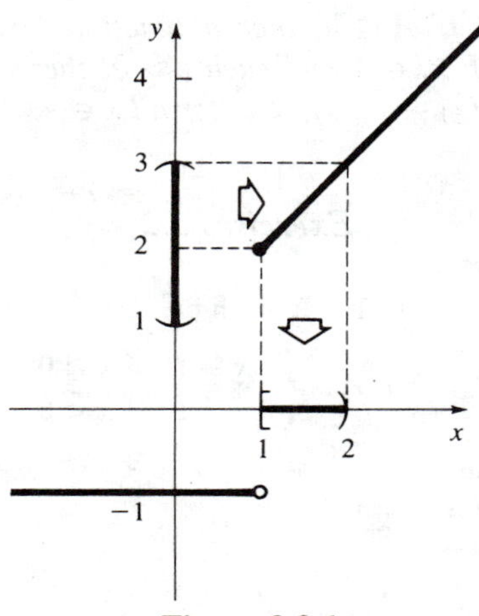

Figure 3.2.1

Example 3.2.11

Let $X = \{a, b, c\}$ and $\mathscr{T} = \{X, \varnothing, \{a\}, \{b\}, \{a, b\}\}$. Let $f: (X, \mathscr{T}) \to (X, \mathscr{T})$ be given by $f(a) = c$, $f(b) = b$, and $f(c) = a$. Then $\{a\}$ is an open set, but $f^{-1}(\{a\}) = \{c\}$ which is not an open set. Thus f is not continuous.

Example 3.2.12

Assume $f: (\mathbb{R}, \mathscr{U}) \to (\mathbb{R}, \mathscr{U})$ is given by $f(x) = x^3$. Let $\mathscr{B}$ be the base for $\mathscr{U}$ consisting of all open intervals. Let $B \in \mathscr{B}$. Then $B = (a, b)$ for some real numbers a and b with $a < b$. Observe that $f^{-1}(B) = (\sqrt[3]{a}, \sqrt[3]{b})$, which is a $\mathscr{U}$-open set. Therefore by Theorem 3.2.9(c) f is continuous.

The following theorems give some of the properties of continuous functions. The proofs are left as exercises.

THEOREM 3.2.13 *Let $(X, \mathscr{T})$, $(Y, \mathscr{S})$, and $(Z, \mathscr{F})$ be topological spaces. If the functions $f: (X, \mathscr{T}) \to (Y, \mathscr{S})$ and $g: (Y, \mathscr{S}) \to (Z, \mathscr{F})$ are continuous, then $g \circ f: (X, \mathscr{T}) \to (Z, \mathscr{F})$ is continuous.*

Recall that if $f: X \to Y$ is a function and $A \subseteq X$, the restriction of f to A is a function from A to Y denoted by $f|_A$ and defined by $f|_A(x) = f(x)$ for all $x \in A$.

THEOREM 3.2.14 *Let $(X, \mathscr{T})$ and $(Y, \mathscr{S})$ be topological spaces and let $A \subseteq X$. If $f: (X, \mathscr{T}) \to (Y, \mathscr{S})$ is continuous, then $f|_A: (A, \mathscr{T}_A) \to (Y, \mathscr{S})$ is continuous.*

We see in the next theorem that if $\mathbb{R}$ has the $\mathscr{U}$ topology, then continuity for a function with range and domain contained in $\mathbb{R}$ is equivalent to the usual ε-δ characterization given in calculus. This is a slight extension of Theorem 2.1.8 and the proof is left as an exercise.

THEOREM 3.2.15 *If $A \subseteq \mathbb{R}$, then a function $f: (A, \mathscr{U}_A) \to (\mathbb{R}, \mathscr{U})$ is continuous* iff *for each $x_0 \in A$ and each $\varepsilon > 0$, there exists $\delta > 0$ such that $|x - x_0| < \delta \Rightarrow |f(x) - f(x_0)| < \varepsilon$, for all $x \in A$.*

Exercises 3.2

1. Determine if the function $f: \mathbb{R} \to \mathbb{R}$ given by

$$f(x) = \begin{cases} -x - 1 & \text{if } x \geqq 0 \\ 1 & \text{if } x < 0 \end{cases}$$

is **(a)** $\mathscr{U}$-$\mathscr{U}$ continuous
(b) $\mathscr{U}$-$\mathscr{H}$ continuous
(c) $\mathscr{U}$-$\mathscr{C}$ continuous
(d) $\mathscr{H}$-$\mathscr{U}$ continuous
(e) $\mathscr{C}$-$\mathscr{C}$ continuous
(f) $\mathscr{C}$-$\mathscr{U}$ continuous
(g) $\mathscr{C}$-$\mathscr{H}$ continuous
(h) $\mathscr{H}$-$\mathscr{H}$ continuous.

2. Determine if the function $f: \mathbb{R} \to \mathbb{R}$ given by

$$f(x) = \begin{cases} -1 & \text{if } x > 0 \\ 1 - x & \text{if } x \leqq 0 \end{cases}$$

is **(a)** $\mathscr{U}$-$\mathscr{U}$ continuous
(b) $\mathscr{U}$-$\mathscr{H}$ continuous
(c) $\mathscr{U}$-$\mathscr{C}$ continuous
(d) $\mathscr{H}$-$\mathscr{U}$ continuous
(e) $\mathscr{C}$-$\mathscr{C}$ continuous
(f) $\mathscr{C}$-$\mathscr{U}$ continuous
(g) $\mathscr{C}$-$\mathscr{H}$ continuous
(h) $\mathscr{H}$-$\mathscr{H}$ continuous.

3. Let $X = \{a, b, c\}$ and $\mathscr{T} = \{X, \varnothing, \{a\}, \{b\}, \{a, b\}\}$. Assume $f: X \to X$ is given by $f(a) = a$, $f(b) = c$, and $f(c) = b$. Determine if f is $\mathscr{T}$-$\mathscr{T}$ continuous.

4. Let $f: \mathbb{R} \to \mathbb{R}$ be defined by $f(x) = x^3 + 1$. Use (c) of Theorem 3.2.9 to prove that f is $\mathscr{U}$-$\mathscr{U}$ continuous.

5. Prove that if $(X, \mathscr{D})$ is a discrete space and $(Y, \mathscr{T})$ is any topological space, then any function $f: (X, \mathscr{D}) \to (Y, \mathscr{T})$ is continuous.

6. Prove that if $(Y, \mathscr{I})$ is an indiscrete space and $(X, \mathscr{T})$ is any topological space, then any function $f: (X, \mathscr{T}) \to (Y, \mathscr{I})$ is continuous.

7. Prove Theorem 3.2.13.

8. Prove that the identity function from $(X, \mathscr{T})$ to $(X, \mathscr{S})$ is continuous iff $\mathscr{T}$ is finer than $\mathscr{S}$.

9. Let $(X, \mathscr{T})$, $(Y, \mathscr{S})$, and $(Y, \mathscr{F})$ be topological spaces and let $f: X \to Y$ be a function. Prove that if f is $\mathscr{T}$-$\mathscr{S}$ continuous and $\mathscr{S}$ is finer than $\mathscr{F}$, then f is $\mathscr{T}$-$\mathscr{F}$ continuous.

10. Let $(X, \mathscr{T})$, $(Y, \mathscr{S})$, and $(Y, \mathscr{F})$ be topological spaces and let $f: X \to Y$ be a function.

Prove that if f is $\mathscr{T}$-$\mathscr{F}$ continuous and $\mathscr{S}$ is finer than $\mathscr{T}$, then f is $\mathscr{S}$-$\mathscr{F}$ continuous.

11. Prove Theorem 3.2.14.

12. Let $(X, \mathscr{T})$ and $(Y, \mathscr{S})$ be topological spaces and let $A \subseteq Y$. Let $f: X \to A$ be a function. Prove that f is $\mathscr{T}$-$\mathscr{S}_A$ continuous iff f is $\mathscr{T}$-$\mathscr{S}$ continuous as a function from X to Y.

13. Which of the following are (i) $\mathscr{U}$-neighborhoods of 1, (ii) $\mathscr{H}$-neighborhoods of 1:

(a) $[0, 1]$ **(b)** $[0, 2]$ **(c)** $[0, 1)$
(d) $[1, 2)$ **(e)** $\{1, 2\}$ **(f)** $(0, 1]$

14. Let $X = \{a, b, c, d\}$ and $\mathscr{T} = \{X, \varnothing, \{a\}, \{b, c\}, \{a, b, c\}\}$. List all $\mathscr{T}$-neighborhoods of

(a) the element a **(b)** the element b.

15. Prove Theorem 3.2.6.

16. Prove Theorem 3.2.7.

17. Prove Theorem 3.2.8.

18. State and prove a theorem that characterizes the boundary of a set in terms of neighborhoods.

19. State and prove a theorem that characterizes the exterior of a set in terms of neighborhoods.

20. Determine the set of all functions $f: \mathbb{R} \to \mathbb{R}$ that are $\mathscr{C}$-$\mathscr{U}$ continuous.

21. Prove Theorem 3.2.15.

3.3 *Homeomorphisms*

In this section conditions are established under which two topological spaces $(X, \mathscr{T})$ and $(Y, \mathscr{S})$ are "topologically identical." Roughly, this means that all properties defined in terms of the open sets of the spaces are the same for both spaces. We shall require that the sets X and Y have the same number of elements. This condition is satisfied by requiring a one-to-one function from X onto Y. We shall also require that the topologies $\mathscr{S}$ and $\mathscr{T}$ have identical structures. Continuity of the one-to-one, onto function meets part of this requirement. However, the following related property is also required.

Definition 3.3.1

Let $(X, \mathscr{T})$ and $(Y, \mathscr{S})$ be topological spaces. A function $f: X \to Y$ is said to be *open* (or $\mathscr{T}$-$\mathscr{S}$ *open*) if for each $\mathscr{T}$-open subset U of X, $f(U)$ is an $\mathscr{S}$-open subset of Y.

Example 3.3.2

Let $f: (\mathbb{R}, \mathscr{U}) \to (\mathbb{R}, \mathscr{H})$ be the identity function. The function f is open since for each $\mathscr{U}$-open set U, $f(U) = U$ which is also $\mathscr{H}$-open. Note that f is not continuous.

Example 3.3.3
Let $X = \{a, b, c\}$, $\mathscr{T} = \{X, \varnothing, \{a\}, \{b\}, \{a, b\}\}$, $Y = \{m, n, p\}$, and $\mathscr{S} = \{Y, \varnothing, \{m\}, \{m, n\}\}$. Let $f: X \to Y$ be given by $f(a) = m$, $f(b) = n$, and $f(c) = p$. Observe that f is not open since $\{b\}$ is a $\mathscr{T}$-open subset of X, but $f(\{b\}) = \{n\}$ which is not an $\mathscr{S}$-open subset of Y.

Example 3.3.4
The function $f: (\mathbb{R}, \mathscr{U}) \to (\mathbb{R}, \mathscr{U})$ given by $f(x) = x^2$ is not open. The details of this fact are left as an exercise.

Definition 3.3.5
Let $(X, \mathscr{T})$ and $(Y, \mathscr{S})$ be topological spaces. A function $f: X \to Y$ is said to be a *homeomorphism* (or a $\mathscr{T}$-$\mathscr{S}$ *homeomorphism*) if f is one-to-one, onto, continuous, and open. If there is a homeomorphism from $(X, \mathscr{T})$ onto $(Y, \mathscr{S})$, then the spaces $(X, \mathscr{T})$ and $(Y, \mathscr{S})$ are said to be *homeomorpic* (usually denoted by $(X, \mathscr{T}) \approx (Y, \mathscr{S})$).

Note the difference in the use of the words "homeomorphism" and "homeomorphic." The word "homeomorphism" refers specifically to a function between the spaces, whereas the term "homeomorphic" describes a relationship between the spaces.

A homeomorphism between two topological spaces establishes a one-to-one correspondence between the open subsets of the two spaces as well as between the points of the two spaces.

If two topological spaces are homeomorphic, then they are *topologically identical.* That is, from a topological viewpoint, the spaces are indistinguishable.

Some of the functions encountered in elementary calculus are homeomorphisms. The exponential function $f(x) = e^x$ is a homeomorphism from $(\mathbb{R}, \mathscr{U})$ onto $(\mathbb{R}^+, \mathscr{U}_{\mathbb{R}^+})$ and the hyperbolic function $g(x) = \tanh(x)$ is a homeomorphism from $(\mathbb{R}, \mathscr{U})$ onto $((-1, 1), \mathscr{U}_{(-1,1)})$. If the domain of $f(x) = 1/x$ is restricted to $\mathbb{R}^+$, then f is a homeomorphism from $(\mathbb{R}^+, \mathscr{U}_{\mathbb{R}^+})$ onto itself.

The following theorem leads to a slightly different characterization of a homeomorphism.

THEOREM 3.3.6 *Let $(X, \mathscr{T})$ and $(Y, \mathscr{S})$ be topological spaces and let $f: X \to Y$ be a one-to-one, onto function. Then f is open* iff $f^{-1}: Y \to X$ *is continuous.*

Proof ($\Rightarrow$) Assume $f: X \to Y$ is one-to-one, onto, and open. Then f^{-1} is a function from Y to X. Let U be a $\mathscr{T}$-open subset of X. Then $(f^{-1})^{-1}(U) = f(U)$ which is an $\mathscr{S}$-open subset of Y. Therefore f^{-1} is continuous.

($\Leftarrow$) Assume $f: X \to Y$ is one-to-one and onto and that $f^{-1}: Y \to X$ is continuous. Let U be a $\mathscr{T}$-open subset of X. Since $f(U) = (f^{-1})^{-1}(U)$, which is an $\mathscr{S}$-open subset of Y, it follows that f is an open function. ■

The proof of the next theorem follows from Theorem 3.3.6 and is left as an exercise.

THEOREM 3.3.7 *Let $(X, \mathcal{T})$ and $(Y, \mathcal{S})$ be topological spaces. A function $f\colon X \to Y$ is a homeomorphism* iff *f is one-to-one, onto, continuous and $f^{-1}\colon Y \to X$ is continuous.*

Example 3.3.8

Let $X = \{a, b, c\}$, $\mathcal{T} = \{X, \varnothing, \{a\}, \{b\}, \{a, b\}\}$, $Y = \{x, y, z\}$, and $\mathcal{S} = \{Y, \varnothing, \{x\}, \{y\}, \{x, y\}\}$. The function $f\colon X \to Y$ given by $f(a) = x$, $f(b) = y$, and $f(c) = z$ is a homeomorphism. The function $g\colon X \to Y$ defined by $g(a) = y$, $g(b) = x$, and $g(c) = z$ is also a homeomorphism. However, the function $h\colon X \to Y$ given by $h(a) = x$, $h(b) = z$, and $h(c) = y$ is not a homeomorphism. Note that $\{b\}$ is a $\mathcal{T}$-open subset of X, but $h(\{b\}) = \{z\}$ which is not $\mathcal{S}$-open in Y. Observe that h is one-to-one and onto.

Example 3.3.9

The spaces $((0, 1), \mathcal{U}_{(0,1)})$ and $((0, 2), \mathcal{U}_{(0,2)})$ are homeomorphic. To see this, show that $f\colon (0, 1) \to (0, 2)$ given by $f(x) = 2x$ is a homeomorphism.

Example 3.3.10

The spaces $(\mathbb{R}, \mathcal{U})$ and $((-\pi/2, \pi/2), \mathcal{U}_{(-\pi/2, \pi/2)})$ are homeomorphic. The function $f\colon (-\pi/2, \pi/2) \to \mathbb{R}$ defined by $f(x) = \tan(x)$ is a homeomorphism.

Definition 3.3.11

A *topological property* is a property P such that if a topological space $(X, \mathcal{T})$ satisfies P then any space homeomorphic to $(X, \mathcal{T})$ also satisfies P.

In order to show that two spaces are not homeomorphic, it is sufficient to find some topological property that one space satisfies and the other does not satisfy. Usually this means finding a property defined in terms of open sets that one space has and the other does not have. For example, one space could have a proper, nonempty open and closed subset and the other space could have no such subset. Note that the number of elements in a space is a topological property.

Example 3.3.12

The spaces $(\mathbb{R}, \mathcal{U})$ and $(\mathbb{R}, \mathcal{H})$ are not homeomorphic. Observe that the set $[0, 1)$ is both $\mathcal{H}$-open and $\mathcal{H}$-closed. However, there is no proper, nonempty subset of $\mathbb{R}$ that is both $\mathcal{U}$-open and $\mathcal{U}$-closed.

Example 3.3.13

Let $X = (0, 2)$ and $Y = (0, 1) \cup (1, 2)$. The spaces $(X, \mathcal{U}_X)$ and $(Y, \mathcal{U}_Y)$ are not homeomorphic. The set $(0, 1)$ is a proper nonempty set that is both $\mathcal{U}_Y$-open and $\mathcal{U}_Y$-closed. The space $(X, \mathcal{U}_X)$ has no such subset, because $(X, \mathcal{U}_X)$ is homeomorphic to $(\mathbb{R}, \mathcal{U})$ (see Example 3.3.12).

Example 3.3.14

Let $X = \{a, b, c\}$, $\mathscr{T} = \{X, \varnothing, \{a, b\}\}$, $Y = \{x, y, z\}$, and $\mathscr{S} = \{Y, \varnothing, \{x\}\}$. The spaces $(X, \mathscr{T})$ and $(Y, \mathscr{S})$ are not homeomorphic. Note that if $f: X \to Y$ were a homeomorphism, then $f(\{a, b\})$ would be a two-element, $\mathscr{S}$-open set.

The following theorems give some of the basic properties of homeomorphisms. The proofs are left as exercises.

THEOREM 3.3.15 *Let $(X, \mathscr{T})$, $(Y, \mathscr{S})$, and $(Z, \mathscr{V})$ be topological spaces. If $f: X \to Y$ and $g: Y \to Z$ are homeomorphisms, then $g \circ f: X \to Z$ is a homeomorphism.*

THEOREM 3.3.16 *Let $(X, \mathscr{T})$ and $(Y, \mathscr{S})$ be topological spaces. If $f: X \to Y$ is a homeomorphism, then $f^{-1}: Y \to X$ is a homeomorphism.*

THEOREM 3.3.17 *Let $(X, \mathscr{T})$ and $(Y, \mathscr{S})$ be topological spaces. If $f: X \to Y$ is a one-to-one, open, continuous function, then $f: (X, \mathscr{T}) \to (f(X), \mathscr{S}_{f(x)})$ is a homeomorphism.*

Exercises 3.3

1. Let $f: (\mathbb{R}, \mathscr{H}) \to (\mathbb{R}, \mathscr{U})$ be the identity function on $\mathbb{R}$. Show that f is not an open function.
2. Let $X = \{a, b, c\}$, $\mathscr{T} = \{X, \varnothing, \{a\}, \{a, b\}\}$, $Y = \{x, y, z\}$, and
$$\mathscr{S} = \{Y, \varnothing, \{x\}, \{y\}, \{x, y\}\}.$$
Show that $g: (X, \mathscr{T}) \to (Y, \mathscr{S})$ given by $g(a) = x$ and $g(b) = g(c) = y$ is an open function. Is g a homeomorphism?
3. Let $\mathscr{S}$ and $\mathscr{T}$ be topologies for a set X. Prove that the identity function from $(X, \mathscr{T})$ onto $(X, \mathscr{S})$ is an open function iff $\mathscr{S}$ is finer than $\mathscr{T}$.
4. Show that the function given in Example 3.3.4 is not open.
5. Let $X = \{a, b, c, d\}$, $\mathscr{T} = \{X, \varnothing, \{a, b\}, \{c, d\}\}$, $Y = \{x, y, z, w\}$, and
$$\mathscr{S} = \{Y, \varnothing, \{x, w\}, \{y, z\}\}.$$
Which of the following functions from $(X, \mathscr{T})$ to $(Y, \mathscr{S})$ are homeomorphisms? If a function is not a homeomorphism, explain why it is not.
   (a) $g(a) = x$, $g(b) = y$, $g(c) = z$, $g(d) = w$
   (b) $g(a) = y$, $g(b) = z$, $g(c) = w$, $g(d) = x$
   (c) $g(a) = g(c) = x$, $g(b) = g(d) = w$
   (d) $g(a) = x$, $g(b) = w$, $g(c) = y$, $g(d) = z$
6. Prove that any topological space is homeomorphic to itself.
7. Show that if $(X, \mathscr{T})$ and $(Y, \mathscr{S})$ are homeomorphic topological spaces and X and Y are finite sets, then X and Y have the same number of elements.

8. Prove that the spaces $((0,1), \mathscr{U}_{(0,1)})$ and $((0,4), \mathscr{U}_{(0,4)})$ are homeomorphic. (See Example 3.3.9.)

9. Let (a,b) and (c,d) be open intervals. Prove that the spaces $((a,b), \mathscr{U}_{(a,b)})$ and $((c,d), \mathscr{U}_{(c,d)})$ are homeomorphic.

10. Prove Theorem 3.3.15.

11. Let (a,b) be an open interval. Show that $((a,b), \mathscr{U}_{(a,b)})$ is homeomorphic to $(\mathbb{R}, \mathscr{U})$. (See Example 3.3.10.)

12. Prove Theorem 3.3.7.

13. Give an example of a set X with topologies $\mathscr{T}$ and $\mathscr{S}$ such that the identity function from $(X, \mathscr{T})$ to $(X, \mathscr{S})$ is continuous but is not a homeomorphism.

14. Let $\mathscr{T} = \{\mathbb{R}, \varnothing, (-\infty, 0), [0, +\infty)\}$. Explain why $(\mathbb{R}, \mathscr{T})$ is not homeomorphic to $(\mathbb{R}, \mathscr{U})$.

15. Let $X = (0,1) \cup (2,4)$ and $Y = (0,4)$. Explain why $(X, \mathscr{U}_X)$ is not homeomorphic to $(Y, \mathscr{U}_Y)$.

16. Let $X = \{a,b,c,d\}$, $\mathscr{T} = \{X, \varnothing, \{a\}, \{a,b\}, \{a,b,c\}\}$, and
$$\mathscr{S} = \{X, \varnothing, \{a,b\}, \{c,d\}\}.$$
Explain why $(X, \mathscr{T})$ is not homeomorphic to $(X, \mathscr{S})$.

17. Let $X = \{a,b,c\}$, $\mathscr{T} = \{X, \varnothing, \{a\}, \{b,c\}, \{a,b,c\}\}$, $Y = \{m,n,p\}$, and
$$\mathscr{S} = \{Y, \varnothing, \{m\}, \{m,n\}\}.$$
Explain why $(X, \mathscr{T})$ and $(Y, \mathscr{S})$ are not homeomorphic.

18. Prove Theorem 3.3.16.

19. Prove Theorem 3.3.17.

20. Recall that $\mathscr{H}$ is the topology on $\mathbb{R}$ with basic open sets of the form $[a,b)$, where $a, b \in \mathbb{R}$. Let $\mathscr{T}$ be the topology on $\mathscr{R}$ with basic open sets of the form $(a,b]$, where $a, b \in \mathbb{R}$. Determine if $(\mathbb{R}, \mathscr{H})$ and $(\mathbb{R}, \mathscr{T})$ are homeomorphic.

21. Determine if there exist topologies $\mathscr{T}$ and $\mathscr{S}$ for $X = \{a,b,c,d\}$ and $Y = \{p,q,r,s,t\}$, respectively, such that $(X, \mathscr{T})$ is homeomorphic to $(Y, \mathscr{S})$.

22. Prove that the space $(\mathbb{R}, \mathscr{U})$ has no proper, nonempty subset that is both open and closed. (This property was used to show that $(\mathbb{R}, \mathscr{U})$ is not homeomorphic to $(\mathbb{R}, \mathscr{H})$.)

3.4 *The Topology of* $\mathbb{R}^n$

In this section the notion of a usual topology for $\mathbb{R}$ is extended to Euclidean n-space, $\mathbb{R}^n$, for any positive integer n. We also take an informal and intuitive look at what it means for subsets of $\mathbb{R}$ or $\mathbb{R}^2$ to be homeomorphic with respect to the usual topology. A rigorous development of these concepts is better suited to an advanced course in topology.

Recall that the usual topology on $\mathbb{R}$ is the topology with a base consisting of all open intervals. The plane, $\mathbb{R}^2$, has an analogous usual topology with a base consisting of all "open disks." Specifically, by an open disk with center (a, b) and radius r, we mean the set

$$D_r(a, b) = \{(x, y) \in \mathbb{R}^2 : \sqrt{(x - a)^2 + (y - b)^2} < r\}.$$

A subset U of $\mathbb{R}^2$ is open with respect to the usual topology provided that for each point $(a, b) \in U$, there is a positive number r such that $D_r(a, b) \subseteq U$. That is, a subset of $\mathbb{R}^2$ is open iff it contains an open disk about each of its points. (See Figure 3.4.1.)

In general for any positive integer $n \geq 3$, $\mathbb{R}^n$ has a usual topology with a base consisting of all "open balls." Specifically, by an open ball with center $(a_1, a_2, \ldots, a_n)$ and radius r, we mean the set

$$B_r(a_1, a_2, \ldots, a_n) = \{(x_1, x_2, \ldots, x_n) \in \mathbb{R}^n : \sqrt{(x_1 - a_1)^2 + (x_2 - a_2)^2 + \cdots + (x_n - a_n)^2} < r\}.$$

A subset U of $\mathbb{R}^n$ is open with respect to the usual topology if, for each point $(a_1, a_2, \ldots, a_n)$ of U, there is a positive number r for which $B_r(a_1, a_2, \ldots, a_n) \subseteq U$. Thus a subset of $\mathbb{R}^n$ is open iff it contains an open ball about each of its points.

In this section we assume that for any positive integer n, $\mathbb{R}^n$ is equipped with its usual topology and that any subset of $\mathbb{R}^n$ has the relative topology induced by the usual topology. Therefore, if X is a subset of $\mathbb{R}^n$, we shall refer to the topological space "X" without listing the topology. This convention will be used frequently in the remaining portion of the book when the specific topology on a set is either understood or unimportant.

There is an especially nice geometric interpretation of homeomorphic subspaces of $\mathbb{R}^n$. If a subspace of $\mathbb{R}^n$ can be deformed into another subspace by such motions as stretching, shrinking, bending, or twisting, but without tearing and without identifying any distinct points, then the two spaces are homeomorphic. We shall refer to these stretching, shrinking, bending, and twisting motions as elastic motions. Technically, this deformation by elastic motions defines a homeomorphism from one space onto the other. However, in this section the geometric interpretation is emphasized rather than the more rigorous functional interpretation. One way to picture this concept is to think of subspaces of $\mathbb{R}^n$ as consisting of perfectly elastic rubber. A space can then be deformed into a second homeomorphic space by stretching, shrinking, twisting, or bending the "rubber" space.

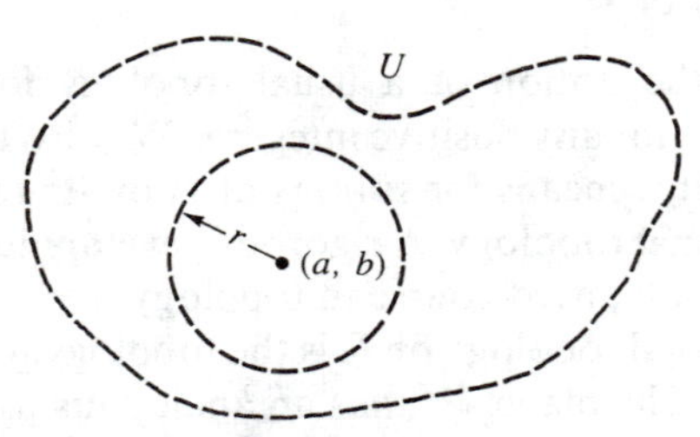

Figure 3.4.1

We have already seen that any two open intervals in $\mathbb{R}$ are homeomorphic. Obviously any given open interval can be deformed into any other open interval by an appropriate stretching or shrinking. Similarly any two line segments (including the endpoints) in $\mathbb{R}^2$ are homeomorphic. All of the spaces in Figure 3.4.2 are homeomorphic, and all of the spaces in Figure 3.4.3 are also homeomorphic. The two spaces in Figure 3.4.4 are not homeomorphic. A line segment cannot be deformed into a circle without identifying two distinct points. Also the circle cannot be deformed into a line segment without cutting the circle. In fact, the number of "pieces" remaining after removing a point is a topological property. Removing the midpoint of a line segment leaves two "pieces," whereas removing any point from a circle leaves only one "piece." Similarly no two of the spaces in Figure 3.4.5 are homeomorphic.

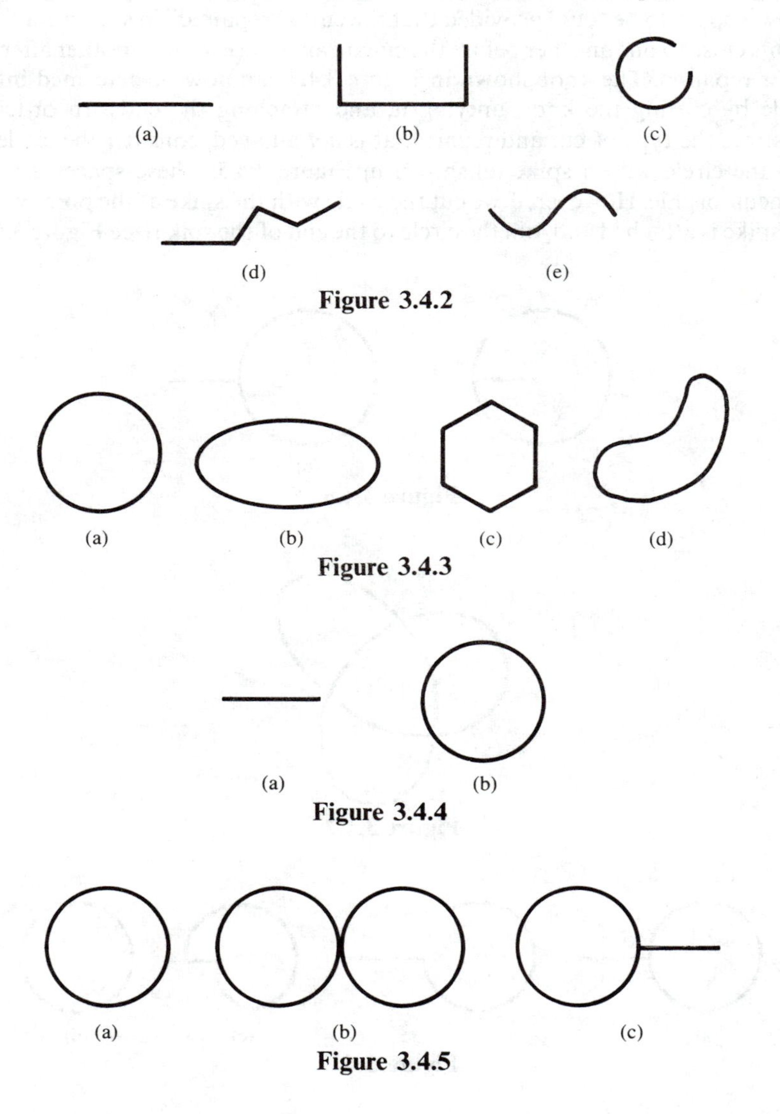

Figure 3.4.2

(a) (b) (c) (d)

Figure 3.4.3

(a) (b)

Figure 3.4.4

(a) (b) (c)

Figure 3.4.5

There is one complication in our geometric interpretation of homeomorphic subspaces of $\mathbb{R}^n$. While it is technically true that two subspaces of $\mathbb{R}^n$ are homeomorphic iff one can be deformed into the other by use of elastic motions, there can be a problem concerning the number of dimensions required for the deformation. Consider the spaces in Figure 3.4.6. Obviously these spaces are homeomorphic. However, the deformation cannot be carried out in two dimensions. There are also examples of homeomorphic spaces where more than three dimensions are required for the deformation. The knot shown in Figure 3.4.7 is homeomorphic to a circle, but the deformation requires four dimensions. Since it is usually very difficult to visualize deformations requiring more than three dimensions, we make the following change in our geometric interpretation of homeomorphic subspaces of $\mathbb{R}^n$. We allow a space to be "cut" provided that the cut is "repaired" in such a way that points close to one another before the cut are also close to one another after the cut is repaired. The knot shown in Figure 3.4.7 can now be deformed into a circle by cutting the knot, untying it, and attaching the ends. In order to illustrate the type of cut and repair that is *not* allowed, consider the circle (a) and the circle with a spike (c) shown in Figure 3.4.5. These spaces are not homeomorphic. However, if we cut the circle with the spike at the point where the spike is attached and join the circle to the end of the spike (see Figure 3.4.8),

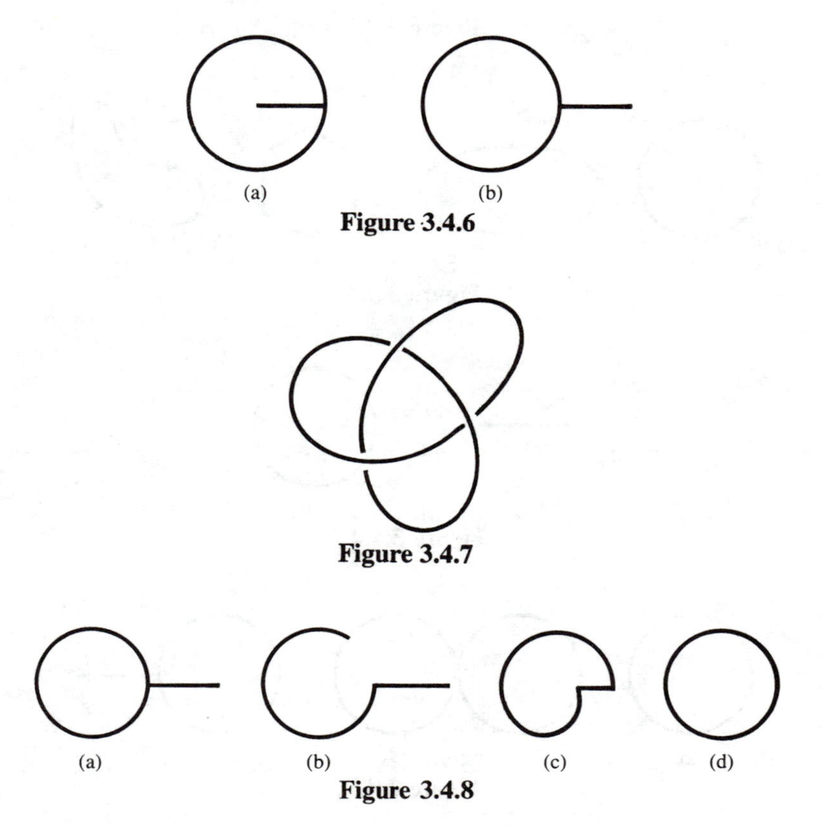

Figure 3.4.6

Figure 3.4.7

Figure 3.4.8

the result is homeomorphic to a circle. Obviously, this cut was not repaired in such a way that points close to one another before the cut were close to one another after the repair.

Exercises 3.4

1. Which of the spaces in Figure 3.4.9 are homeomorphic to a circle? Which spaces are homeomorphic to a line segment?
2. Partition the spaces in Figure 3.4.10 into mutually disjoint collections of homeomorphic spaces such that, if two spaces belong to different collections, then they are not homeomorphic.
3. Write the letters of the alphabet as block capital letters. Partition these "spaces" into mutually disjoint collections of homeomorphic spaces such that any two spaces from different collections are not homeomorphic.
4. List the numerals one through nine as displayed by a calculator. Partition these "spaces" into mutually disjoint collections of homeomorphic spaces such that any two spaces from different collections are not homeomorphic.

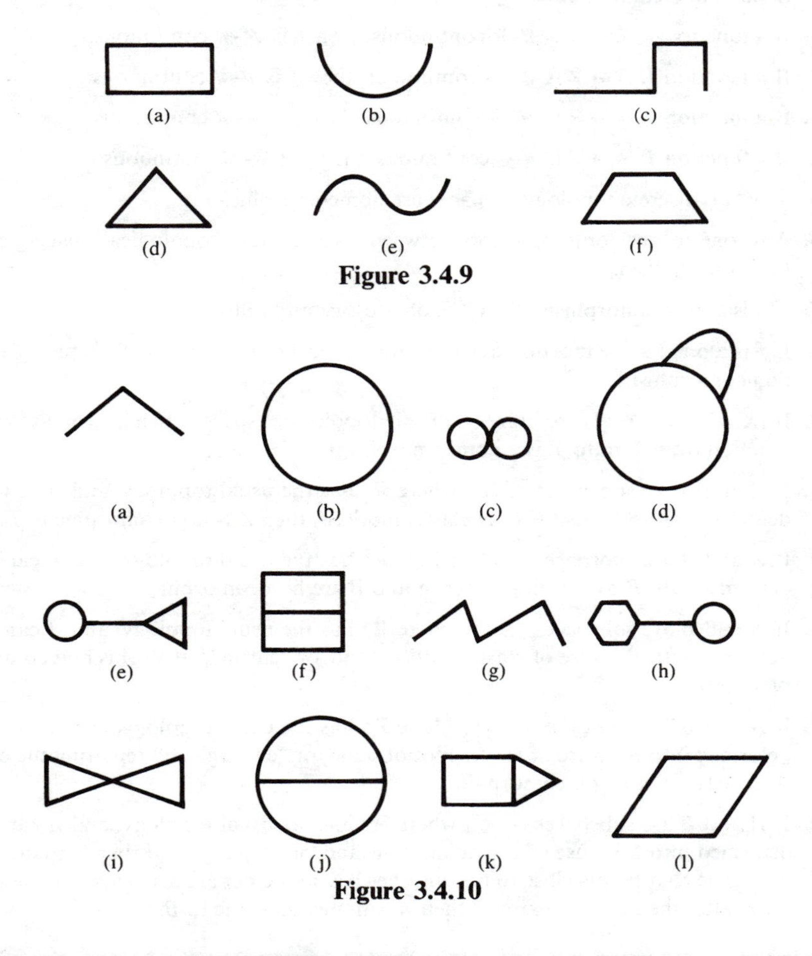

Figure 3.4.9

Figure 3.4.10

Review Exercises 3

Mark each of the following statements true or false. Briefly explain each true statement and find a counterexample for each false statement.

1. Let $(X, \mathscr{T})$ be a topological space with $A \subseteq X$ and $U \subseteq A$. If U is $\mathscr{T}_A$-open, then U is $\mathscr{T}$-open.
2. Let $(X, \mathscr{T})$ be a topological space with $A \subseteq X$ and $U \subseteq A$. If U is $\mathscr{T}$-open, then U is $\mathscr{T}_A$-open.
3. If $(X, \mathscr{T})$ is a topological space with $A \subseteq X$, then $\mathscr{T}_A \subseteq \mathscr{T}$.
4. If $(X, \mathscr{T})$ is a topological space and A is an open subset of X, then $\mathscr{T}_A \subseteq \mathscr{T}$.
5. Any open set is a neighborhood of each of its points.
6. Every neighborhood of a point in a topological space is also an open set.
7. Every constant function is continuous regardless of the topologies on the domain and codomain.
8. The identity function is always continuous regardless of the topologies on the domain and codomain.
9. If a function $f\colon \mathbb{R} \to \mathbb{R}$ is $\mathscr{U}$-$\mathscr{U}$ continuous, then f is $\mathscr{H}$-$\mathscr{U}$ continuous.
10. If a function $f\colon \mathbb{R} \to \mathbb{R}$ is $\mathscr{U}$-$\mathscr{H}$ continuous, then f is $\mathscr{U}$-$\mathscr{U}$ continuous.
11. If a function $f\colon \mathbb{R} \to \mathbb{R}$ is $\mathscr{H}$-$\mathscr{U}$ continuous, then f is $\mathscr{U}$-$\mathscr{U}$ continuous.
12. If a function $f\colon \mathbb{R} \to \mathbb{R}$ is $\mathscr{C}$-$\mathscr{U}$ continuous, then f is $\mathscr{U}$-$\mathscr{U}$ continuous.
13. Any two discrete topological spaces are homeomorphic.
14. Any one-to-one, onto function between two discrete topological spaces is a homeomorphism.
15. If f is a homeomorphism, then f is one-to-one and onto.
16. If f is a one-to-one function from one topological space onto another, then f is a homeomorphism.
17. If $(X, \mathscr{T})$ and $(Y, \mathscr{S})$ are homeomorphic topological spaces, then any one-to-one function from X onto Y is a homeomorphism.
18. If A and B are subspaces of $\mathbb{R}^2$, where $\mathbb{R}^2$ has the usual topology, and A can be deformed into B by use of only elastic motions, then A is homeomorphic to B.
19. If A and B are subspaces of $\mathbb{R}^2$, where $\mathbb{R}^2$ has the usual topology, and A can be deformed into B by "cutting," then A and B are homeomorphic.
20. If A and B are subspaces of $\mathbb{R}^2$, where $\mathbb{R}^2$ has the usual topology, and A can be deformed into B by use of elastic motions and/or "cutting," then A is homeomorphic to B.
21. If A and B are subspaces of $\mathbb{R}^2$, where $\mathbb{R}^2$ has the usual topology, and A can be deformed into B by use of elastic motions and/or "cutting" and repairing the cut, then A and B are homeomorphic.
22. If A and B are subspaces of $\mathbb{R}^2$, where $\mathbb{R}^2$ has the usual topology, and A can be deformed into B by use of elastic motions and/or "cutting" and then "repairing" the cut so that points close to one another before the cut are also close to one another after the cut is "repaired." then A is homeomorphic to B.

4

Product Spaces

4.1 *Products of Two Topological Spaces*

For two topological spaces $(X, \mathscr{T})$ and $(Y, \mathscr{S})$, there is a natural way to define a topology on the product set $X \times Y$. The obvious method is to consider all possible products of open subsets of X with open subsets of Y. The problem is that this collection of sets is not closed under the operation of union and hence does not form a topology. However, as we see in the following theorem, the collection does form a base for a topology.

THEOREM 4.1.1 *Let $(X, \mathscr{T})$ and $(Y, \mathscr{S})$ be topological spaces. The collection $\mathscr{B} = \{U \times V : U \in \mathscr{T}, V \in \mathscr{S}\}$ is a base for a topology on the product set $X \times Y$.*

Proof We must show that parts (a) and (b) of Theorem 2.5.7 hold. To see that (a) holds, note that $X \times Y \in \mathscr{B}$ and hence clearly $X \times Y = \bigcup\{B : B \in \mathscr{B}\}$.

To prove (b), let $B_1, B_2 \in \mathscr{B}$ and let $(x, y) \in B_1 \cap B_2$. There exist open subsets U_1 and U_2 of X and open subsets V_1 and V_2 of Y for which $B_1 = U_1 \times V_1$ and $B_2 = U_2 \times V_2$. Then $B_1 \cap B_2 = (U_1 \times V_1) \cap (U_2 \times V_2)$. By Exercise 1, $(U_1 \times V_1) \cap (U_2 \times V_2) = (U_1 \cap U_2) \times (V_1 \cap V_2)$. Obviously $U_1 \cap U_2$ and $V_1 \cap V_2$ are open subsets of X and Y, respectively, and hence $(U_1 \cap U_2) \times (V_1 \cap V_2) \in \mathscr{B}$. Thus if we let $B_3 = (U_1 \cap U_2) \times (V_1 \cap V_2)$, then $B_3 \in \mathscr{B}$ and $(x, y) \in B_3 \subseteq B_1 \cap B_2$.

Therefore $\mathscr{B}$ is a base for a topology on $X \times Y$. ■

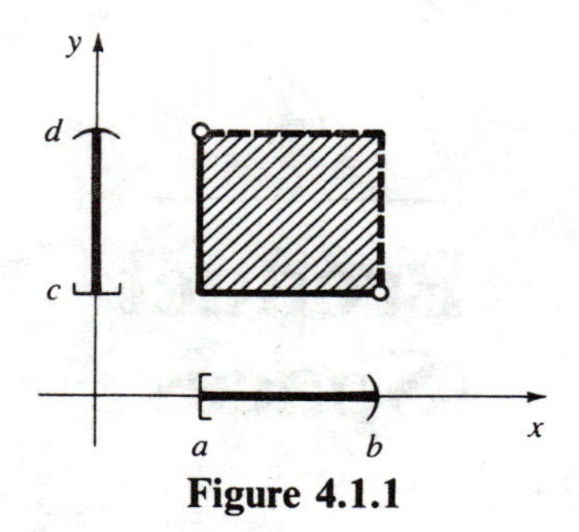

Figure 4.1.1

Definition 4.1.2

Let $(X, \mathscr{T})$ and $(Y, \mathscr{S})$ be topological spaces. The topology generated by the base $\mathscr{B}$ given in Theorem 4.1.1 is called the *product topology* for $X \times Y$. The set $X \times Y$ together with the product topology is called a *product space.*

Note that a subset W of a product space $X \times Y$ is open iff, for each $(x, y) \in W$, there are open subsets U and V of X and Y, respectively, for which $(x, y) \in U \times V \subseteq W$.

Example 4.1.3

Let $\mathbb{R}$ have the $\mathscr{U}$ topology. A typical open set in the product space $\mathbb{R} \times \mathbb{R}$ (or $\mathbb{R}^2$) is an "open rectangle" of the form $(a, b) \times (c, d)$ where (a, b) and (c, d) are open intervals. The product topology on $\mathbb{R} \times \mathbb{R}$ is the same as the topology generated by open disks.

Example 4.1.4

Let $\mathbb{R}$ have the $\mathscr{H}$ topology. A typical open set in the product space $\mathbb{R} \times \mathbb{R}$ is a "half-open rectangle" of the form $[a, b) \times [c, d)$ (see Figure 4.1.1). This product topology on $\mathbb{R} \times \mathbb{R}$, introduced by R. H. Sorgenfrey in 1947, is known as Sorgenfrey's half-open square topology.

Example 4.1.5

Let $X = \mathbb{R}$ with the $\mathscr{U}$ topology and let $Y = \mathbb{R}$ with the $\mathscr{C}$ topology. A typical open set in the product space $X \times Y$ is a set of the form $(b, c) \times (a, +\infty)$. (See Figure 4.1.2.)

Example 4.1.6

If both X and Y have the discrete topology, then the product topology for $X \times Y$ is also the discrete topology. If both X and Y have the indiscrete topology, then the product topology for $X \times Y$ is the indiscrete topology.

Example 4.1.7

Let $X = \{a, b, c\}$, $\mathscr{T} = \{X, \varnothing, \{a\}, \{a, b\}\}$, $Y = \{m, n\}$, and

$$\mathscr{S} = \{Y, \varnothing, \{m\}\}.$$

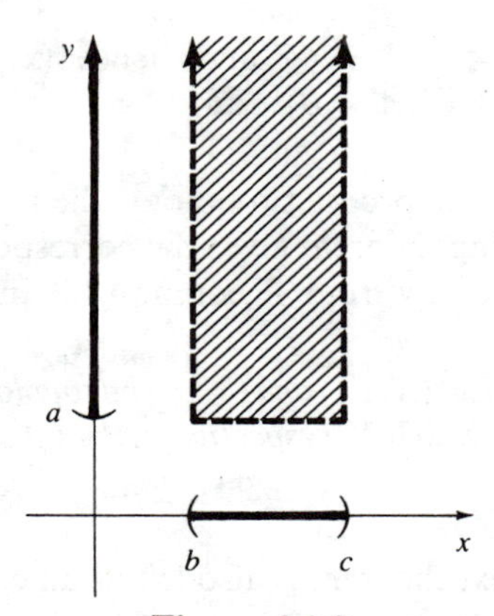

Figure 4.1.2

The base $\mathscr{B}$ for the product topology on $X \times Y$ described in Theorem 4.1.1 is the set

$$\mathscr{B} = \{X \times Y, \varnothing, X \times \{m\}, \{a\} \times \{m\}, \{a\} \times Y, \{a,b\} \times Y, \{a,b\} \times \{m\}\}.$$

In order to form a base for a product space $X \times Y$, it is sufficient to use only products of basic open subsets of X and Y.

THEOREM 4.1.8 *Let $(X, \mathscr{T}_1)$ and $(Y, \mathscr{T}_2)$ be topological spaces. If $\mathscr{B}_1$ and $\mathscr{B}_2$ are bases for $\mathscr{T}_1$ and $\mathscr{T}_2$, respectively, then the collection $\mathbb{B} = \{B_1 \times B_2 : B_1 \in \mathscr{B}_1, B_2 \in \mathscr{B}_2\}$ is a base for the product topology on $X \times Y$.*

The proof is left as an exercise.

THEOREM 4.1.9 *Let $(X, \mathscr{T})$ and $(Y, \mathscr{S})$ be topological spaces with $A \subseteq X$ and $B \subseteq Y$. Then $\mathrm{Cl}(A \times B) = \mathrm{Cl}(A) \times \mathrm{Cl}(B)$.*

Proof Let $(x, y) \in \mathrm{Cl}(A \times B)$. Let U be any open subset of X containing x and let V be any open subset of Y containing y. Then $U \times V$ is an open subset of the product space $X \times Y$ containing (x, y). Since $(x, y) \in \mathrm{Cl}(A \times B)$, it follows that $(U \times V) \cap (A \times B) \neq \varnothing$. Because

$$(U \times V) \cap (A \times B) = (U \cap A) \times (V \cap B)$$

(see Exercise 1), we have that $U \cap A \neq \varnothing$ and $V \cap B \neq \varnothing$. Thus $x \in \mathrm{Cl}(A)$ and $y \in \mathrm{Cl}(B)$. Therefore $(x, y) \in \mathrm{Cl}(A) \times \mathrm{Cl}(B)$ and hence $\mathrm{Cl}(A \times B) \subseteq \mathrm{Cl}(A) \times \mathrm{Cl}(B)$.

To see that $\mathrm{Cl}(A) \times \mathrm{Cl}(B) \subseteq \mathrm{Cl}(A \times B)$, let $(x, y) \in \mathrm{Cl}(A) \times \mathrm{Cl}(B)$. Let W be any open subset of the product space $X \times Y$ containing (x, y). Then there is a basic open set of the form $U \times V$, where U is open in X and V is open in Y, for which $(x, y) \in U \times Y \subseteq W$. Therefore $x \in U$ and $y \in V$. Since $x \in \mathrm{Cl}(A)$ and $y \in \mathrm{Cl}(B)$, we have that $U \cap A \neq \varnothing$ and $V \cap B \neq \varnothing$. Thus $(U \times V) \cap (A \times B) = (U \cap A) \times (V \cap B) \neq \varnothing$. It follows that $W \cap (A \times B) \neq \varnothing$ and hence $(x, y) \in \mathrm{Cl}(A \times B)$. Therefore $\mathrm{Cl}(A) \times \mathrm{Cl}(B) \subseteq \mathrm{Cl}(A \times B)$. ■

By the definition of the product topology, the product of any two open sets is open. The following theorem gives the corresponding result for closed sets. The proof follows easily from Theorem 4.1.9 and is left as an exercise.

THEOREM 4.1.10 *Let $(X, \mathscr{T})$ and $(Y, \mathscr{S})$ be topological spaces. If A and B are closed subsets of X and Y, respectively, then $A \times B$ is a closed subset of $X \times Y$.*

The proof of the next theorem is also left as an exercise.

THEOREM 4.1.11 *Let $(X, \mathscr{T})$ and $(Y, \mathscr{S})$ be topological spaces with $A \subseteq X$ and $B \subseteq Y$. Then $\mathrm{Int}(A \times B) = \mathrm{Int}(A) \times \mathrm{Int}(B)$.*

Exercises 4.1

1. Let X and Y be sets with U and A subsets of X and V and B subsets of Y. Prove that $(A \times B) \cap (U \times V) = (A \cap U) \times (B \cap V)$.
2. Let $X = \mathbb{R}$ have the $\mathscr{U}$ topology and $Y = \mathbb{R}$ have the $\mathscr{H}$ topology. Describe a typical open subset of
   (a) the product space $X \times Y$
   (b) the product space $Y \times X$.
3. Let $X = \mathbb{R}$ with the $\mathscr{C}$ topology. Describe a typical open subset of the product space $X \times X$.
4. Let $X = \mathbb{R}$ have the discrete topology, $\mathscr{D}$, and $Y = \mathbb{R}$ have the indiscrete topology, $\mathscr{I}$. Describe the product topology on $X \times Y$.
5. Let $X = \mathbb{R}$ with the topology $\mathscr{T} = \{U \subseteq X : 1 \in U \text{ or } U = \varnothing\}$. Let $Y = \mathbb{R}$ with the topology $\mathscr{S} = \{U \subseteq Y : 2 \in U \text{ or } U = \varnothing\}$. Describe the product topology on $X \times Y$.
6. Let $X = \{a, b, c\}$, $\mathscr{T} = \{X, \varnothing, \{a\}, \{b\}, \{a, b\}\}$, $Y = \{p, q\}$, and $\mathscr{S} = \{Y, \varnothing, \{q\}\}$. List the sets in the base for the product topology on $X \times Y$ given in Theorem 4.1.1.
7. Prove Theorem 4.1.8.
8. Let $\mathbb{R}$ have the usual topology. Describe a subset of $\mathbb{R}^2 = \mathbb{R} \times \mathbb{R}$ that is open in the product space but that is not a product of open subsets of $\mathbb{R}$.
9. Prove Theorem 4.1.10.
10. Prove Theorem 4.1.11.
11. Let $(X, \mathscr{T})$ and $(Y, \mathscr{S})$ be topological spaces with $A \subseteq X$ and $B \subseteq Y$. Prove or disprove the following: $\mathrm{Ext}(A \times B) = \mathrm{Ext}(A) \times \mathrm{Ext}(B)$.

4.2 Finite Products and Projections

In this section the concept of the product of two topological spaces is extended to the product of any finite number of spaces. Properties of certain functions defined on product spaces are investigated also.

If $(X_1, \mathscr{T}_1), (X_2, \mathscr{T}_2), \ldots, (X_n, \mathscr{T}_n)$ are topological spaces, then the product set $X_1 \times X_2 \times \cdots \times X_n$ is the collection of all ordered n-tuples of the form $(x_1, x_2, \ldots, x_n)$ where $x_i \in X_i$ for each $i \in \{1, 2, \ldots, n\}$. The following theorem gives a base for a topology on the product set.

THEOREM 4.2.1 *Let $(X_1, \mathscr{T}_1), (X_2, \mathscr{T}_2), \ldots, (X_n, \mathscr{T}_n)$ be topological spaces. The collection $\mathscr{B} = \{U_1 \times U_2 \times \cdots \times U_n : U_i \in \mathscr{T}_i$ for $i = 1, 2, \ldots, n\}$ is a base for a topology on the set $X_1 \times X_2 \times \cdots \times X_n$.*

The proof is analogous to that of Theorem 4.1.1 and is left as an exercise.

Definition 4.2.2

Let $(X_1, \mathscr{T}_1), (X_2, \mathscr{T}_2), \ldots, (X_n, \mathscr{T}_n)$ be topological spaces. The topology generated by the base $\mathscr{B}$ given in Theorem 4.2.1 is called the *product topology* for $X_1 \times X_2 \times \cdots \times X_n$. The set $X_1 \times X_2 \times \cdots \times X_n$ together with the product topology is called a *product space.*

The next theorem is similar to Theorem 4.1.8.

THEOREM 4.2.3 *Let $(X_1, \mathscr{T}_1), (X_2, \mathscr{T}_2), \ldots, (X_n, \mathscr{T}_n)$ be topological spaces. If $\mathscr{B}_i$ is a base for $\mathscr{T}_i$ for each $i \in \{1, 2, \ldots, n\}$, then the collection $\mathbb{B} = \{B_1 \times B_2 \times \cdots \times B_n : B_i \in \mathscr{B}_i$ for $i = 1, 2, \ldots, n\}$ is a base for the product space $X_1 \times X_2 \times \cdots \times X_n$.*

The proof is straightforward and is left as an exercise.

Example 4.2.4

Let $\mathbb{R}$ have the $\mathscr{U}$ topology and let $\mathscr{B}$ be the base for $\mathscr{U}$ consisting of all open intervals. For $\mathbb{R}^3 = \mathbb{R} \times \mathbb{R} \times \mathbb{R}$ a typical set in the base given in Theorem 4.2.3 is an "open box" of the form $I_1 \times I_2 \times I_3$ where I_1, I_2, and I_3 are open intervals. The product topology on $\mathbb{R}^3$ with the open boxes as a base is the same as the topology on $\mathbb{R}^3$ with the open balls as a base. That is, the collection of open balls and the collection of open boxes are different bases for the same topology.

Example 4.2.5

Let $\mathbb{R}$ have the $\mathscr{H}$ topology. A typical basic open set in the product space $\mathbb{R}^3$ is a "half-open box" of the form $[a, b) \times [c, d) \times [e, f)$ where $a, b, c, d, e, f \in \mathbb{R}$.

The functions given in the following definition play an important role in the theory of product spaces.

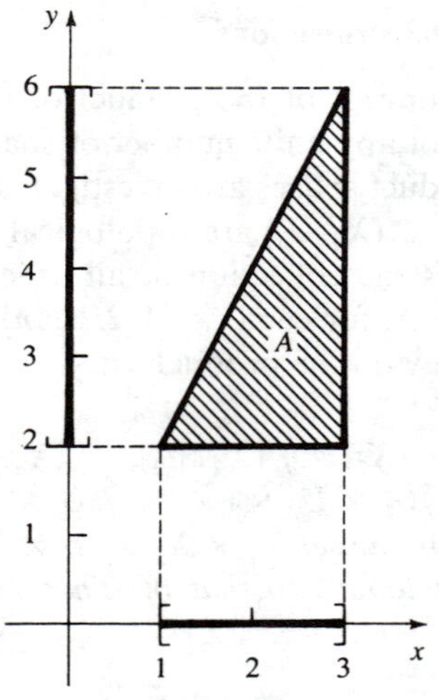

Figure 4.2.1

Definition 4.2.6

Let $X_1, X_2, \ldots, X_n$ be sets. For each $i \in \{1, 2, \ldots, n\}$ the function $p_i: X_1 \times X_2 \times \cdots \times X_n \to X_i$ given by $p_i(x_1, x_2, \ldots, x_n) = x_i$ is called the *ith projection* or the *ith projection function.*

Example 4.2.7

Let $X_1 = X_2 = \mathbb{R}$ and let $p_1: X_1 \times X_2 \to X_1$ and $p_2: X_1 \times X_2 \to X_2$ be the projections. Let $A = \{(x, y): 1 \leqq x \leqq 3, 2 \leqq y \leqq 2x\}$ (see Figure 4.2.1). Then $p_1(A) = [1, 3]$ and $p_2(A) = [2, 6]$.

The following two rather technical results will simplify our proofs involving projections.

LEMMA 4.2.8 *Let $X_1, X_2, \ldots, X_n$ be sets and let $i \in \{1, 2, \ldots, n\}$. If $V_i \subseteq X_i$, then $p_i^{-1}(V_i) = X_1 \times X_2 \times \cdots \times X_{i-1} \times V_i \times X_{i+1} \times \cdots \times X_n$.*

Proof Let $(x_1, x_2, \ldots, x_n) \in X_1 \times X_2 \times \cdots \times X_n$. Then $(x_1, x_2, \ldots, x_n) \in p_i^{-1}(V_i)$ iff $P_i(x_1, x_2, \ldots, x_n) \in V_i$ iff $x_i \in V_i$ iff $(x_1, x_2, \ldots, x_n) \in X_1 \times X_2 \times \cdots \times X_{i-1} \times V_i \times X_{i+1} \times \cdots \times X_n$. ■

LEMMA 4.2.9 *Let $X_1, X_2, \ldots, X_n$ be sets and let $V_i \subseteq X_i$ for each $i \in \{1, 2, \ldots, n\}$. Then $\bigcap\{p_i^{-1}(V_i): i = 1, 2, \ldots, n\} = V_1 \times V_2 \times \cdots \times V_n$.*

Proof Let $(x_1, x_2, \ldots, x_n) \in X_1 \times X_2 \times \cdots \times X_n$. Then $(x_1, x_2, \ldots, x_n) \in \bigcap\{p_i^{-1}(V_i): i = 1, 2, \ldots, n\}$ iff $(x_1, x_2, \ldots, x_n) \in p_i^{-1}(V_i)$ for each i iff $p(x_1, x_2, \ldots, x_n) \in V_i$ for each i iff $x_i \in V_i$ for each i iff $(x_1, x_2, \ldots, x_n) \in V_1 \times V_2 \times \cdots \times V_n$. ■

Example 4.2.10

Let $X_1 = X_2 = \mathbb{R}$ and let $p_1: X_1 \times X_2 \to X_1$ and $p_2: X_1 \times X_2 \to X_2$ be the projections. Let $V_1 = [1,2] \subseteq X_1$ and $V_2 = [3,4] \subseteq X_2$. Then $p_1^{-1}(V_1) = [1,2] \times \mathbb{R}$ and $p_2^{-1}(V_2) = \mathbb{R} \times [3,4]$. Also $p_1^{-1}(V_1) \cap p_2^{-1}(V_2) = [1,2] \times [3,4]$ (See Figure 4.2.2).

The following theorems give some of the basic properties of projections.

THEOREM 4.2.11 *Let $(X_1, \mathscr{T}_1)$, $(X_2, \mathscr{T}_2), \ldots, (X_n, \mathscr{T}_n)$ be topological spaces. For each $j \in \{1, 2, \ldots, n\}$, the projection $p_j\colon X_1 \times X_2 \times \cdots \times X_n \to X_j$ is an open, continuous function.*

Proof Let $j \in \{1, 2, \ldots, n\}$. In order to show that p_j is an open function, it is sufficient to show that the image of each basic open set is open. Let U be a basic open set of the form $U = U_1 \times U_2 \times \cdots \times U_n$, where U_i is an open subset of X_i for each i. Clearly $p_j(U) = U_j$ which is obviously an open subset of X_j.

To show that p_j is continuous, let V_j be an open subset of X_j. By Lemma 4.2.8 $p_j^{-1}(V_j) = X_1 \times X_2 \times \cdots \times X_{j-1} \times V_j \times X_{j+1} \times \cdots \times X_n$ which is clearly an open set in the product space. ■

THEOREM 4.2.12 *Let $(X_1, \mathscr{T}_1)$, $(X_2, \mathscr{T}_2), \ldots, (X_n, \mathscr{T}_n)$ be topological spaces. The product topology is the coarsest topology on $X_1 \times X_2 \times \cdots \times X_n$ for which all the projections are continuous.*

Proof Let $\mathscr{F}$ be any topology on the set $X_1 \times X_2 \times \cdots \times X_n$ for which all the projections are continuous. We must show that the product topology is contained in $\mathscr{F}$. It is sufficient to show that any basic open set in the product topology is also in $\mathscr{F}$. Let W be a basic open subset of the product space of the form $W = U_1 \times U_2 \times \cdots \times U_n$ where $U_i \in \mathscr{T}_i$ for each i. By Lemma 4.2.9 $W = \bigcap\{p_i^{-1}(U_i) : i = 1, 2, \ldots, n\}$. Since p_i is $\mathscr{F}$-$\mathscr{T}_i$ continuous for each i, it follows that $p_i^{-1}(U_i)$ is $\mathscr{F}$-open for each i. Then

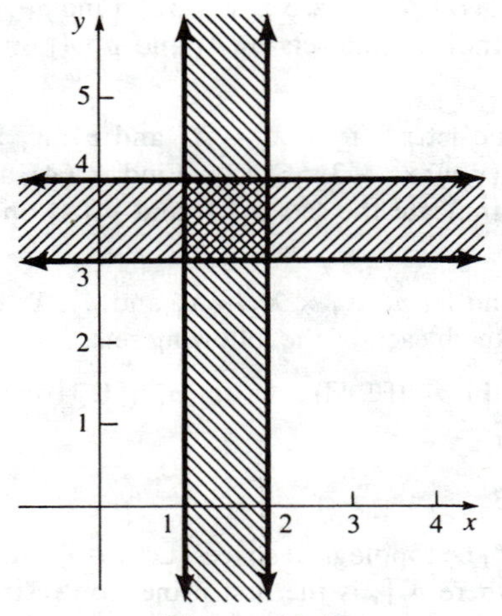

Figure 4.2.2

$\bigcap\{p_i^{-1}(U_i): i = 1, 2, \ldots, n\}$ is $\mathscr{F}$-open since it is a finite intersection of $\mathscr{F}$-open sets. Thus the product topology is contained in $\mathscr{F}$. ■

THEOREM 4.2.13 *Let $(Y_1, \mathscr{T}_1)$, $(Y_2, \mathscr{T}_2), \ldots, (Y_n, \mathscr{T}_n)$, and $(X, \mathscr{S})$ be topological spaces. A function $f: X \to Y_1 \times Y_2 \times \cdots \times Y_n$ is continuous iff $p_i \circ f$ is continuous for each $i \in \{1, 2, \ldots, n\}$, where p_i is the ith projection.*

Proof ($\Rightarrow$) Assume f is continuous. By Theorem 4.2.11 each projection, p_i, is continuous. Since the composition of continuous functions is continuous, it follows that $p_i \circ f$ is continuous for each i.

($\Leftarrow$) Assume $p_i \circ f$ is continuous for each i. In order to show that f is continuous, it is sufficient to show that the inverse image of each basic open set is open. Let W be a basic open set in the product space of the form $W = V_1 \times V_2 \times \cdots \times V_n$, where V_i is an open subset of Y_i for each i. By Lemma 4.2.9 $W = \bigcap\{p_i^{-1}(V_i): i = 1, 2, \ldots, n\}$. Hence $f^{-1}(W) = f^{-1}(\bigcap\{p_i^{-1}(V_i): i = 1, 2, \ldots, n\}) = \bigcap\{f^{-1}(p_i^{-1}(V_i)): i = 1, 2, \ldots, n\} = \bigcap\{(p_i \circ f)^{-1}(V_i): i = 1, 2, \ldots, n\}$ which is a finite intersection of $\mathscr{S}$-open sets and hence is $\mathscr{S}$-open. ■

Exercises 4.2

1. Let $\mathbb{R}$ have the $\mathscr{H}$ topology. Sketch a "half-open box" in the product topology for $\mathbb{R}^3 = \mathbb{R} \times \mathbb{R} \times \mathbb{R}$.

2. Let $X = \mathbb{R}$ with the topology $\mathscr{T} = \{U \subseteq X : U = \varnothing \text{ or } 1 \in U\}$. Describe a typical open subset of the product space $X^4 = X \times X \times X \times X$.

3. Let $X = \mathbb{R}$ with the topology $\mathscr{T} = \{U \subseteq X : U = \varnothing \text{ or } 1 \in U\}$, $Y = \mathbb{R}$ with the topology $\mathscr{S} = \{U \subseteq Y : U = Y \text{ or } 2 \notin U\}$, and $Z = \mathbb{R}$ with the topology $\mathscr{F} = \{U \subseteq Z : U = \varnothing \text{ or } 2 \in U\}$. Describe a typical open subset of the product space $X \times Y \times Z$.

4. Let $X_1 = X_2 = \mathbb{R}$ and let $p_1: X_1 \times X_2 \to X_1$ and $p_2: X_1 \times X_2 \to X_2$ be the projections. Let $A = \{(x, y): 1 \leqq x \leqq 2, 3 \leqq y \leqq 3x\}$. Find $p_1(A)$ and $p_2(A)$. Sketch the set A in the plane. Indicate the sets $p_1(A)$ and $p_2(A)$ on the x-axis and y-axis, respectively.

5. Let $X_1 = X_2 = \mathbb{R}$ and let $p_1: X_1 \times X_2 \to X_1$ and $p_2: X_1 \times X_2 \to X_2$ be the projections. Let $A = \{(x, y): x^2 + 3y^2 \leqq 12\}$. Find $p_1(A)$ and $p_2(A)$. Sketch the set A in the plane. Indicate the sets $p_1(A)$ and $p_2(A)$ on the x-axis and y-axis, respectively.

6. Let $X_1 = X_2 = \mathbb{R}$, and let $p_1: X_1 \times X_2 \to X_1$ and $p_2: X_1 \times X_2 \to X_2$ be the projections. Find and sketch each of the following sets:

   (a) $p_1^{-1}([1, 3])$ **(b)** $p_2^{-1}([2, 4])$ **(c)** $p_1^{-1}([1, 3]) \cap p_2^{-1}([2, 4])$

7. Prove Theorem 4.2.1.

8. Prove Theorem 4.2.3.

9. Let $(X, \mathscr{T})$ and $(Y, \mathscr{S})$ be topological spaces. Let $a \in Y$ and let $A = X \times \{a\}$. Show that $p_1|_A: A \to X$, where $p_1|_A$ is the first projection restricted to the set A, is a homeomorphism.

10. Let $(X_1, \mathcal{T}_1), (X_2, \mathcal{T}_2), \ldots, (X_n, \mathcal{T}_n)$ be topological spaces. Let F_i be a closed subset of X_i for each $i \in \{1, 2, \ldots, n\}$. Prove that $F_1 \times F_2 \times \cdots \times F_n$ is a closed subset of the product space $X_1 \times X_2 \times \cdots \times X_n$.

11. Let $(X_1, \mathcal{T}_1)$, $(X_2, \mathcal{T}_2), \ldots, (X_n, \mathcal{T}_n)$ be topological spaces. Let $U_i \subseteq X_i$ for each $i \in \{1, 2, \ldots, n\}$. Prove that $\mathrm{Cl}\{U_1) \times \mathrm{Cl}(U_2) \times \cdots \times \mathrm{Cl}(U_n) = \mathrm{Cl}(U_1 \times U_2 \times \cdots \times U_n)$.

12. Let $(X_1, \mathcal{T}_1)$, $(X_2, \mathcal{T}_2), \ldots, (X_n, \mathcal{T}_n)$ be topological spaces. Let $U_i \subseteq X_i$ for each $i \in \{1, 2, \ldots, n\}$. Prove that

$$\mathrm{Int}(U_1) \times \mathrm{Int}(U_2) \times \cdots \times \mathrm{Int}(U_n) = \mathrm{Int}(U_1 \times U_2 \times \cdots \times U_n).$$

4.3* *Infinite Products*

We now extend the concept of the product of a finite number of topological spaces to the product of an infinite number of spaces. In order to accomplish this, we must first redefine the product of a finite number of spaces.

Let X_1, $X_2, \ldots, X_n$ be a finite number of sets. For each ordered n-tuple $(x_1, x_2, \ldots, x_n)$ in the product set $X_1 \times X_2 \times \cdots \times X_n$, define the function $x\colon \{1, 2, \ldots, n\} \to \bigcup\{X_i : i = 1, 2, \ldots, n\}$ by $x(i) = x_i$ for each i. This establishes a one-to-one correspondence between the n-tuples of the form $(x_1, x_2, \ldots, x_n)$ and the functions $x\colon \{1, 2, \ldots, n\} \to \bigcup\{X_i : i = 1, 2, \ldots, n\}$ such that $x(i) \in X_i$ for each i. Thus we have the following alternate definition of a product of a finite number of sets.

Definition 4.3.1

(Alternate definition of the product of a finite number of sets.) The product of a finite number of sets $X_1, X_2, \ldots, X_n$ is denoted by $X_1 \times X_2 \times \cdots \times X_n$ and is defined to be the set of all functions

$$x\colon \{1, 2, \ldots, n\} \to \bigcup\{X_i : i = 1, 2, \ldots, n\}$$

such that $x(i) \in X_i$ for each i.

Figure 4.3.1 is an illustration of the alternate definition for $\mathbb{R}^3$.

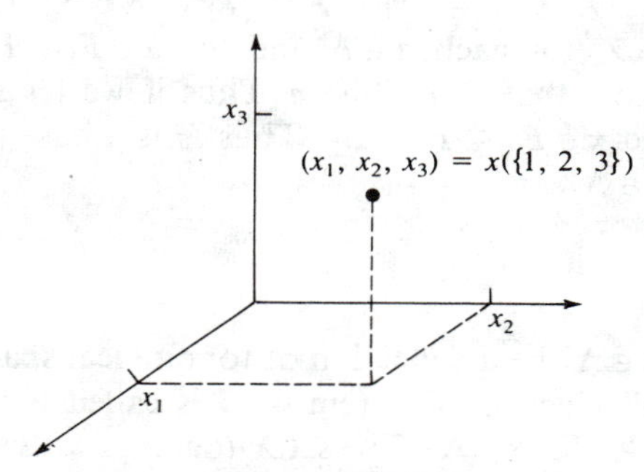

Figure 4.3.1

Next consider the product of an arbitrary collection of sets $\{X_\alpha : \alpha \in \Lambda\}$. The index set Λ can be any nonempty set, finite or infinite. For example, Λ can be the set of real numbers. In this case the usual notion of an n-tuple obviously cannot be used. However, Definition 4.3.1 easily extends to the product of an arbitrary collection of sets.

Definition 4.3.2

The product of the collection $\{X_\alpha : \alpha \in \Lambda\}$ of sets is denoted by $\mathbf{X}\{X_\alpha : \alpha \in \Lambda\}$ and is defined to be the set of all functions

$$x: \Lambda \to \bigcup\{X_\alpha : \alpha \in \Lambda\}$$

such that $x(\alpha) \in X_\alpha$ for each $\alpha \in \Lambda$. We shall denote $x(\alpha)$ by x_α and refer to x_α as the αth coordinate of x.

Our next task is to build a topology for the product of an arbitrary collection of topological spaces, $\{X_\alpha : \alpha \in \Lambda\}$. The obvious approach is to consider all products of the form $\mathbf{X}\{U_\alpha : \alpha \in \Lambda\}$ where U_α is an open subset of X_α for each $\alpha \in \Lambda$. This collection does indeed form a base for a topology called the box topology. However, there are complications with this topology. For example, Theorem 4.2.13 would not extend to infinite products if the box topology were used. In order to avoid these problems, we require U_α to equal X_α except for a finite number of α. As we see in the next theorem, this collection also forms a base for a topology.

THEOREM 4.3.3 *Let $\{(X_\alpha, \mathscr{T}_\alpha) : \alpha \in \Lambda\}$ be a collection of topological spaces. The collection $\mathscr{B}$ of all sets of the form $\mathbf{X}\{U_\alpha : \alpha \in \Lambda\}$, where $U_\alpha \in \mathscr{T}_\alpha$ for all $\alpha \in \Lambda$ and $U_\alpha = X_\alpha$ except for a finite number of α is a base for a topology on the product set $\mathbf{X}\{X_\alpha : \alpha \in \Lambda\}$.*

Proof We must show that parts (a) and (b) of Theorem 2.5.7 hold. Obviously (a) holds since $\mathbf{X}\{X_\alpha : \alpha \in \Lambda\} \in \mathscr{B}$.

To see that (b) holds, let $B_1, B_2 \in \mathscr{B}$ and $x \in B_1 \cap B_2$. Then there are open subsets U_α and V_α of X_α for each $\alpha \in \Lambda$ such that $B_1 = \mathbf{X}\{U_\alpha : \alpha \in \Lambda\}$ and $B_2 = \mathbf{X}\{V_\alpha : \alpha \in \Lambda\}$ and finite subsets F_1 and F_2 of Λ for which $U_\alpha = X_\alpha$ if $\alpha \notin F_1$ and $V_\alpha = X_\alpha$ if $\alpha \notin F_2$. Note that $B_1 \cap B_2 = (\mathbf{X}\{U_\alpha : \alpha \in \Lambda\}) \cap (\mathbf{X}\{V_\alpha : \alpha \in \Lambda\}) = \mathbf{X}\{U_\alpha \cap V_\alpha : \alpha \in \Lambda\}$ (see Exercise 10). Since $U_\alpha \cap V_\alpha \in \mathscr{T}_\alpha$ for each $\alpha \in \Lambda$ and for $\alpha \notin F_1 \cup F_2$ $U_\alpha \cap V_\alpha = X_\alpha \cap X_\alpha = X_\alpha$, it follows that $B_1 \cap B_2 \in \mathscr{B}$. Thus if we let $B_3 = B_1 \cap B_2$, we have $B_3 \in \mathscr{B}$ and $x \in B_3 \subseteq B_1 \cap B_2$. Thus $\mathscr{B}$ is a base for a topology on the set $\mathbf{X}\{X_\alpha : \alpha \in \Lambda\}$. ■

Definition 4.3.4

Let $\{(X_\alpha, \mathscr{T}_\alpha) : \alpha \in \Lambda\}$ be a collection of topological spaces. The topology with the base $\mathscr{B}$ given in Theorem 4.3.3 is called the *product topology* for the set $X = \mathbf{X}\{X_\alpha : \alpha \in \Lambda\}$. The set X together with the product topology is called a *product space*.

The product topology is the only topology we shall consider for a product set. The following definition extends the notion of a projection to the product of an infinite number of sets.

Definition 4.3.5
Let $\{X_\alpha : \alpha \in \Lambda\}$ be a collection of sets. For each $\beta \in \Lambda$ the function $p_\beta : \mathbf{X}\{X_\alpha : \alpha \in \Lambda\} \to X_\beta$, defined by $p_\beta(x) = x_\beta$ where $x_\beta = x(\beta)$, is the *βth projection* or the *βth projection function.*

The product topology can be characterized in terms of the projections. The next definition is required for this characterization.

Definition 4.3.6
Let $(X, \mathscr{T})$ be a topological space. A collection $\mathscr{S}$ of subsets of X is said to be a *subbase* for $\mathscr{T}$ provided that the collection of all finite intersections of sets in $\mathscr{S}$ forms a base for $\mathscr{T}$.

Example 4.3.7
The collection $\mathscr{S}$ of all intervals of the form $(a, +\infty)$ or $(-\infty, b)$ is a subbase for the $\mathscr{U}$ topology on $\mathbb{R}$.

Example 4.3.8
The collection $\mathscr{S}$ of all intervals of the form $[a, +\infty)$ or $(-\infty, b)$ is a subbase for the $\mathscr{H}$ topology on $\mathbb{R}$.

Example 4.3.9
Let X be a set with more than two elements. The collection $\mathscr{S}$ of all subsets of X containing two elements is a subbase for the $\mathscr{D}$ topology on X.

THEOREM 4.3.10 *Let $\{(X_\alpha, \mathscr{T}_\alpha) : \alpha \in \Lambda\}$ be a collection of topological spaces. The collection $\mathscr{S} = \{p_\alpha^{-1}(U_\alpha) : U_\alpha \in \mathscr{T}_\alpha, \alpha \in \Lambda\}$ is a subbase for the product topology on the set $\mathbf{X}\{X_\alpha : \alpha \in \Lambda\}$.*

Proof Let $\mathscr{B}$ be the base for $\mathbf{X}\{X_\alpha : \alpha \in \Lambda\}$ given in Theorem 4.3.3. Let $U = \mathbf{X}\{U_\alpha : \alpha \in \Lambda\}$ be a member of $\mathscr{B}$, where $U_\alpha \in \mathscr{T}_\alpha$ for each α and there is a finite subset F of Λ for which $U_\alpha = X_\alpha$ for $\alpha \notin F$. We shall show the following:

$$U = \bigcap\{p^{-1}(U_\alpha) : \alpha \in F\}.$$

Let $x \in U$. Then $x \colon \Lambda \to \bigcup\{U_\alpha : \alpha \in \Lambda\}$ such that $x(\alpha) = x_\alpha \in U_\alpha$ for each $\alpha \in \Lambda$. Then for each $\alpha \in F$, $p_\alpha(x) = x_\alpha \in U_\alpha$ and hence $x \in p_\alpha^{-1}(U_\alpha)$. Therefore $U \subseteq \bigcap\{p_\alpha^{-1}(U_\alpha) : \alpha \in F\}$.

Let $x \in \bigcap\{p_\alpha^{-1}(U_\alpha) : \alpha \in F\}$. Then for each $\alpha \in F$, $x \in p_\alpha^{-1}(U_\alpha)$ and thus $x_\alpha = p_\alpha(x) \in U_\alpha$. For $\alpha \in \Lambda - F$, $U_\alpha = X_\alpha$ and hence clearly $p_\alpha(x) = x_\alpha \in U_\alpha$. Then since $p_\alpha(x) = x_\alpha \in U_\alpha$ for all α, it follows that $x \in \mathbf{X}\{U_\alpha : \alpha \in \Lambda\}$. Therefore $\bigcap\{p_\alpha^{-1}(U_\alpha) : \alpha \in F\} \subseteq U$. ∎

The product topology could have been defined as the topology with the subbase given in Theorem 4.3.10. The following theorems extend some of our results for finite products to infinite products.

THEOREM 4.3.11 *Let $\{(X_\alpha, \mathscr{T}_\alpha): \alpha \in \Lambda\}$ be a collection of topological spaces. For each $\beta \in \Lambda$ the projection $p_\beta: \mathbf{X}\{X_\alpha: \alpha \in \Lambda\} \to X_\beta$ is an open, continuous function.*

Proof Let $\beta \in \Lambda$ and let $V_\beta \in \mathscr{T}_\beta$. Since $p_\beta^{-1}(V_\beta)$ is a member of the subbase for the product topology on $\mathbf{X}\{X_\alpha: \alpha \in \Lambda\}$ given in Theorem 4.3.10, it follows that $p_\beta^{-1}(V_\beta)$ is open in the product space. Hence p_β is continuous.

In order to show that p_β is open, it is sufficient to show that the image of each basic open set is open. Let U be a basic open subset of the product space of the form $\mathbf{X}\{U_\alpha: \alpha \in \Lambda\}$ where $U_\alpha \in \mathscr{T}_\alpha$ for each α and $U_\alpha = X_\alpha$ except for a finite number of α. We shall show that $p_\beta(U) = U_\beta$. From the definition of the projection p_β, it is clear that $p_\beta(U) = \{p_\beta(x): x \in U\} = \{x_\beta: x \in U\} \subseteq U_\beta$. To see that $U_\beta \subseteq p_\beta(U)$, let $x_\beta \in U_\beta$. Then for each $\alpha \in \Lambda$ different from β let x_α be any element of U_α. Let x be the element of U defined by $x(\alpha) = x_\alpha$ for each $\alpha \in \Lambda$. By construction of x, it is clear that $P_\beta(x) = x_\beta$ and hence $x_\beta \in p_\beta(U)$. Therefore $U_\beta \subseteq p_\beta(U)$. Thus $p_\beta(U) = U_\beta$ and it follows that p_β is an open function. ■

THEOREM 4.3.12 *Let $\{(X_\alpha, \mathscr{T}_\alpha): \alpha \in \Lambda\}$ be a collection of topological spaces. The product topology is the coarsest topology on $\mathbf{X}\{X_\alpha: \alpha \in \Lambda\}$ for which all the projections are continuous.*

The proof follows easily from Theorem 4.3.10 and is left as an exercise.

LEMMA 4.3.13 *Let $(X, \mathscr{T})$ and $(Y, \mathscr{F})$ be topological spaces and let $\mathscr{S}$ be a subbase for $\mathscr{F}$. Then a function $f: X \to Y$ is continuous iff $f^{-1}(S)$ is open for each $S \in \mathscr{S}$.*

This proof is also left as an exercise.

THEOREM 4.3.14 *Let $\{(Y_\alpha, \mathscr{T}_\alpha): \alpha \in \Lambda\}$ be a collection of topological spaces and let $(X, \mathscr{F})$ be a topological space. A function*

$$f: X \to \mathbf{X}\{Y_\alpha: \alpha \in \Lambda\}$$

is continuous iff $p_\alpha \circ f: X \to Y_\alpha$ is continuous for each $\alpha \in \Lambda$.

Proof ($\Rightarrow$) Assume f is continuous. Then $p_\alpha \circ f$ is continuous for each $\alpha \in \Lambda$ because the composition of continuous functions is continuous by Theorem 3.2.13.

($\Leftarrow$) Assume $p_\alpha \circ f$ is continuous for each $\alpha \in \Lambda$. By Lemma 4.3.13, in order to show that f is continuous, it is sufficient to show that the inverse image of any subbasic open set is open. Let S be a subbasic open set of the product space of the form $p_\alpha^{-1}(V_\alpha)$, where $\alpha \in \Lambda$ and $V_\alpha \in \mathscr{T}_\alpha$. Then $f^{-1}(S) = f^{-1}(p_\alpha^{-1}(V_\alpha)) = (p_\alpha \circ f)^{-1}(V_\alpha)$ which is $\mathscr{F}$-open since $p_\alpha \circ f$ is continuous. ■

This section is concluded with several examples which illustrate the nature of the product topology on the product of an infinite number of sets.

Example 4.3.15

Let $\Lambda = \mathbb{Z}^+$ and for each $i \in \Lambda$ let $X_i = \mathbb{R}$ and $\mathcal{T}_i = \mathcal{U}$. Let $X = \mathbf{X}\{X_i : i \in \Lambda\}$.

(a) $\mathbf{X}\{U_i : i \in \Lambda\}$, where U_1 is the open interval $(0, 1)$ and for $i \neq 1$, $U_i = X_i$ is an open subset of the product space X.

(b) $\mathbf{X}\{V_i : i \in \Lambda\}$, where for each $i \in \{2, 4, 6, 8, 10\}$ V_i is the open interval $(-i, i)$ and otherwise $V_i = X_i$, is an open subset of the product space X.

(c) $\mathbf{X}\{W_i : i \in \Lambda\}$, where W_i is the open interval $(0, 1)$ for each $i \in \Lambda$, is neither open nor closed in the product space X.

(d) $\mathbf{X}\{F_i : i \in \Lambda\}$, where $F_i = [0, 1]$ for each $i \in \Lambda$, is a closed subset of the product space X.

(e) $\mathbf{X}\{G_i : i \in \Lambda\}$, where G_i is the open interval $(0, 1)$ for each odd integer i and $G_i = X_i$ for each even integer, is neither open nor closed in the product space X.

(f) Let $x \in X$ such that $x(i) = x_i = 1$ for each $i \in \Lambda$. The set $\{x\}$ is a closed subset of the product space X. To see this, let $y \in X - \{x\}$. Then there exists $j \in \Lambda$ for which $y_j \neq 1$. Let U_j be an open interval containing y_j but not 1. For $i \neq j$, let $U_i = X_i$. The set $U = \mathbf{X}\{U_i : i \in \Lambda\}$ is open in the product space X and $y \in U \subseteq X - \{x\}$. (This method can be used to show that any finite subset of the product space X is closed.)

Example 4.3.16

Let $\Lambda = \mathbb{Z}^+$. For each $i \in \Lambda$, let $X_i = \mathbb{R}$ and $\mathcal{T}_i = \mathcal{D}$. The product topology on $\mathbf{X}\{X_i : i \in \Lambda\}$ is not the discrete topology. For example, if $U_i = \{i\}$ for each $i \in \Lambda$, then $\mathbf{X}\{U_i : i \in \Lambda\}$ is not open in the product space $\mathbf{X}\{X_i : i \in \Lambda\}$.

Exercises 4.3

1. Show that the collection $\mathcal{S}$ given in Example 4.3.7 is a subbase for the $\mathcal{U}$ topology. That is, show that each open interval can be expressed as a finite intersection of sets in $\mathcal{S}$.
2. Show that the collection $\mathcal{S}$ given in Example 4.3.8 is a subbase for the $\mathcal{H}$ topology. That is, show that each interval of the form $[a, b)$ can be expressed as a finite intersection of sets in $\mathcal{S}$.
3. Show that the collection $\mathcal{S}$ given in Example 4.3.9 is a subbase for the $\mathcal{D}$ topology.
4. Prove Lemma 4.3.13.
5. Let $\{(X_\alpha, \mathcal{T}_\alpha) : \alpha \in \Lambda\}$ be a collection of topological spaces. Let W be a nonempty open subset of the product space $\mathbf{X}\{X_\alpha : \alpha \in \Lambda\}$. Prove that there exist $\beta \in \Lambda$ such

that for each $a \in X_\beta$, there is an $x \in W$ with $x_\beta = a$. That is, show that there is some $\beta \in \Lambda$ for which the βth coordinate of elements of W is unrestricted. What can be said about the number of $\beta \in \Lambda$ for which the βth coordinate of the elements of W is unrestricted?

6. Let $\Lambda = [0,9]$ and for each $\alpha \in \Lambda$ let $X_\alpha = \mathbb{R}$ and let $\mathscr{T}_\alpha = \mathscr{D}$. Which of the following are open subsets of the product space $\mathbf{X}\{X_\alpha : \alpha \in \Lambda\}$? If a set is not open, explain why it is not.

(a) $\mathbf{X}\{U_\alpha : \alpha \in \Lambda\}$, where $U_\alpha = \{\alpha, \alpha + 1\}$ for each $\alpha \in \Lambda$
(b) $\mathbf{X}\{V_\alpha : \alpha \in \Lambda\}$, where $V_\alpha = \{\alpha, \alpha + 1\}$ for $\alpha \in \{1,2,7\}$ and otherwise $V_\alpha = X_\alpha$
(c) $\mathbf{X}\{W_\alpha : \alpha \in \Lambda\}$, where $W_\alpha = \{\alpha, \alpha + 1\}$ for $\alpha \in [0,1]$ and otherwise $W_\alpha = X_\alpha$.

7. Let $\Lambda = [0,1)$ and for each $\alpha \in \Lambda$ let $X_\alpha = \mathbb{R}$ and $\mathscr{T}_\alpha = \mathscr{I}$. Which of the following sets are open in the product space $\mathbf{X}\{X_\alpha : \alpha \in \Lambda\}$? If a set is not open, explain why it is not.

(a) $\mathbf{X}\{U_\alpha : \alpha \in \Lambda\}$, where $U_\alpha = \varnothing$ for each $\alpha \in \Lambda$
(b) $\mathbf{X}\{V_\alpha : \alpha \in \Lambda\}$, where $V_0 = \varnothing$ and, for $\alpha \neq 0$, $V_\alpha = \{\alpha\}$
(c) $\mathbf{X}\{W_\alpha : \alpha \in \Lambda\}$, where $W_0 = \{0\}$ and, for $\alpha \neq 0$, $W_\alpha = X_\alpha$.

8. Let $\Lambda = \mathbb{Z}^+$ and for each $i \in \Lambda$, let $X_i = \mathbb{R}$ and let $\mathscr{T}_i = \mathscr{H}$. Which of the following are open subsets of the product space $\mathbf{X}\{X_i : i \in \Lambda\}$? If a set is not open, explain why it is not.

(a) $\mathbf{X}\{U_i : i \in \Lambda\}$, where $U_i = [0,1)$ for each $i \in \Lambda$
(b) $\mathbf{X}\{V_i : i \in \Lambda\}$, where $V_i = [0,1)$ if i is an odd integer and $V_i = X_i$ if i is an even integer
(c) $\mathbf{X}\{W_i : i \in \Lambda\}$, where $W_5 = [0,5)$ and for $i \neq 5$, $W_i = X_i$.

9. Let $\Lambda = \mathbb{Z}^+$ and for each $i \in \Lambda$, let $X_i = \mathbb{R}$ and $\mathscr{T}_i = \mathscr{H}$. Use a method similar to that of Example 4.3.15(f) to show that each of the following sets is a closed subset of the product space $X = \mathbf{X}\{X_i : i \in \Lambda\}$.

(a) $\{x\}$, where $x(i) = x_i = 3$ for each $i \in \Lambda$
(b) $\{x\}$, where $x(i) = x_i = i$ for each $i \in \Lambda$
(c) $\{x, y\}$, where $x(i) = x_i = 2$ for each $i \in \Lambda$ and $y(i) = y_i = 4$ for each $i \in \Lambda$.

10. Let $\{X_\alpha : \alpha \in \Lambda\}$ be a collection of sets. For each $\alpha \in \Lambda$, let U_α and V_α be subsets of X_α. Prove that $(\mathbf{X}\{U_\alpha : \alpha \in \Lambda\}) \cap (\mathbf{X}\{V_\alpha : \alpha \in \Lambda\}) = \mathbf{X}\{U_\alpha \cap V_\alpha : \alpha \in \Lambda\}$.

11. Prove Theorem 4.3.12.

12. Let $W = \mathbf{X}\{W_i : i \in \Lambda\}$ be as in Example 4.3.15(c). Show that if x is the element of X defined by $x(i) = x_i = 0$ for each $i \in \Lambda$, then x is a limit point of W. Use this fact to show that W is not closed.

13. Let $\{(X_\alpha, \mathscr{T}_\alpha) : \alpha \in \Lambda\}$ be a collection of topological spaces. Assume Λ is an infinite set. Let $A_\alpha \subseteq X_\alpha$ for each $\alpha \in \Lambda$. Prove or disprove the following:

$$\text{Int}(\mathbf{X}\{A_\alpha : \alpha \in \Lambda\}) = \mathbf{X}\{\text{Int}(A_\alpha) : \alpha \in \Lambda\}.$$

14. Prove that the set $\mathbf{X}\{F_i : i \in \Lambda\}$ given in Example 4.3.15(d) is a closed subset of the product space X.

15. Let $\{(X_\alpha, \mathscr{T}_\alpha) : \alpha \in \Lambda\}$ be a collection of topological spaces. Let $A_\alpha \subseteq X_\alpha$ for each $\alpha \in \Lambda$. Prove that $\text{Cl}(\mathbf{X}\{A_\alpha : \alpha \in \Lambda\}) = \mathbf{X}\{\text{Cl}(A_\alpha) : \alpha \in \Lambda\}$.

16. Let $\{(X_\alpha, \mathscr{T}_\alpha) : \alpha \in \Lambda\}$ be a collection of topological spaces. Let F_α be a closed subset of X_α for each $\alpha \in \Lambda$. Prove that $\mathbf{X}\{F_\alpha : \alpha \in \Lambda\}$ is a closed subset of the product space $\mathbf{X}\{X_\alpha : \alpha \in \Lambda\}$.

4.4 *Continuity of Algebraic Operations on* ℝ

The product topology provides the opportunity to study the continuity of the algebraic operations on the real numbers. The seemingly diverse notions of algebra and topology come together to exhibit uniquely complementary properties in the study of the continuity of these algebraic operations.

In particular, we are considering the following functions:

$a\colon \mathbb{R} \times \mathbb{R} \to \mathbb{R}$, defined by $a(p, q) = p + q$

$m\colon \mathbb{R} \times \mathbb{R} \to \mathbb{R}$, defined by $m(p, q) = pq$

$i_a\colon \mathbb{R} \to \mathbb{R}$ (additive inverse), defined by $i_a(p) = -p$

$i_m\colon \mathbb{R} - \{0\} \to \mathbb{R}$ (multiplicative inverse), defined by $i_m(p) = 1/p$.

Example 4.4.1

$a^{-1}(\{3\}) = \{(x, y) : y = -x + 3\}$, and

$$a^{-1}((-1, 2)) = \{(x, y) : -x - 1 < y < -x + 2\}$$

(see Figure 4.4.1).

Example 4.4.2

$m^{-1}(\{-1\}) = \{(x, y) : y = -1/x\}$, and

$$m^{-1}([1, 2)) = \{(x, y) : 1/x \leqq y < 2/x\}$$

(see Figure 4.4.2).

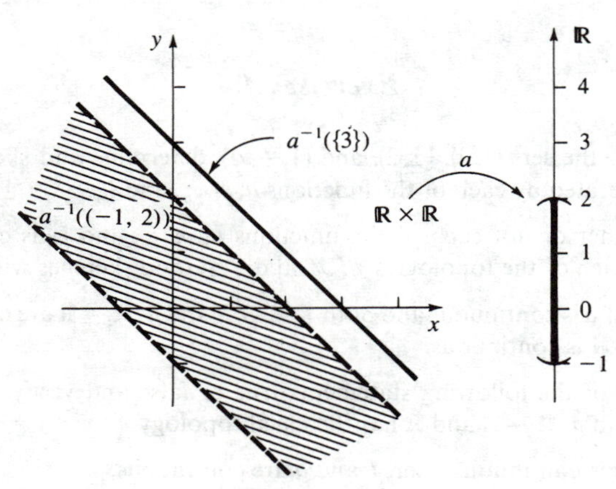

Figure 4.4.1

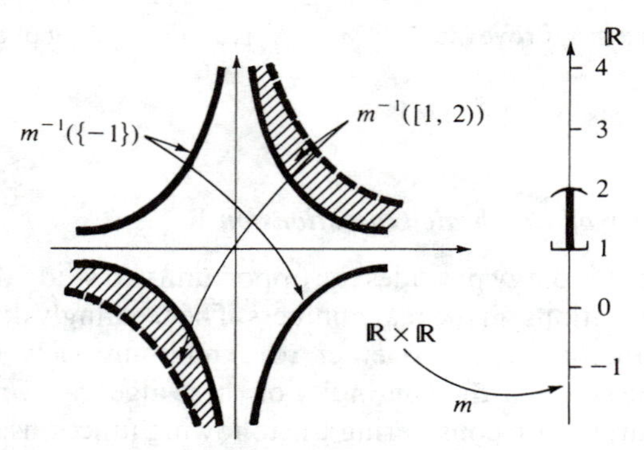

Figure 4.4.2

Example 4.4.3
$i_a^{-1}((3, +\infty)) = (-\infty, -3)$, and $i_m^{-1}([2, 4)) = (1/4, 1/2]$.

In the exercises that follow this section, we shall see that the continuity of the algebraic functions a, m, i_a, and i_m, where $\mathbb{R}$ has the usual topology, is the result of the careful development of an appropriate "compatible" topology for $\mathbb{R}$. In fact, minor tinkering with the usual topology, such as adjoining the left endpoint to each of the basic open sets (generating the $\mathscr{H}$ topology) results in the discontinuity of m, i_a, and i_m. The continuity of a, m, i_a, and i_m has appeared before in theorems from calculus stating that the sums and products of continuous functions are continuous. To illustrate the important role that the usual topology on $\mathbb{R}$ has played in the foundations of the calculus, we consider the function $f: \mathbb{R} \to \mathbb{R}$, defined by $f(x) = x$, which is continuous with respect to either of the topologies $\mathscr{H}$ or $\mathscr{C}$. However, the product of f with itself, $(f \cdot f)(x) = x^2$, is not continuous with respect to either of the topologies $\mathscr{H}$ or $\mathscr{C}$.

This has been an early glimpse into some exciting topics for further study in the area of topological algebra.

Exercises 4.4

1. For each of the sets $(1, 2)$, $[1, 2)$, and $(1, +\infty)$, determine and sketch the inverse images generated by each of the functions a, m, i_a, and i_m.
2. Verify whether or not each of the functions a, m, i_a, and i_m is continuous with respect to each of the topologies $\mathscr{U}$, $\mathscr{H}$, and $\mathscr{C}$. (Good sketches will suffice.)
3. Prove that if a is continuous and both $f: \mathbb{R} \to \mathbb{R}$ and $g: \mathbb{R} \to \mathbb{R}$ are continuous, then $f + g: \mathbb{R} \to \mathbb{R}$ is continuous.
4. Mark each of the following statements true or false and verify. In this exercise $f: \mathbb{R} \to \mathbb{R}$ and $g: \mathbb{R} \to \mathbb{R}$ and $\mathbb{R}$ has the usual topology.

   (a) If $f + g$ is continuous, then f and g are continuous.

(**b**) If fg is continuous, then f and g are continuous.
(**c**) If f and g are continuous, then $f + g$ is discontinuous.
(**d**) If f and g are discontinuous, then fg is discontinuous.
(**e**) If f is continuous and g is discontinuous, then $f + g$ is discontinuous.

5. Let $f: \mathbb{R} \to \mathbb{R}$ be defined by $f(x) = x$. Verify that the product of this function with itself, $(f \cdot f)(x) = x^2$, is not continuous, if $\mathbb{R}$ has either of the topologies $\mathscr{H}$ or $\mathscr{C}$.

Review Exercises 4

Mark each of the following statements true or false. Briefly explain each true statement and find a counterexample for each false statement.

1. If $(X, \mathscr{T})$ and $(Y, \mathscr{F})$ are topological spaces, then the collection

$$\{U \times V : U \in \mathscr{T}, V \in \mathscr{F}\}$$

is a topology on $X \times Y$.

2. If $(X, \mathscr{T})$ and $(Y, \mathscr{F})$ are topological spaces, then the collection

$$\{U \times V : U \in \mathscr{T}, V \in \mathscr{F}\}$$

is a base for a topology on $X \times Y$.

3. If X has the discrete topology and Y has the discrete topology then the product topology for $X \times Y$ is the discrete topology.

***4.** If $\{(X_\alpha, \mathscr{T}_\alpha) : \alpha \in \Lambda\}$ is a collection of discrete topological spaces and Λ is infinite, then the product topology for $\mathbf{X}\{X_\alpha : \alpha \in \Lambda\}$ is the discrete topology.

5. If X has the discrete topology and Y has the indiscrete topology, then the product topology for $X \times Y$ is the indiscrete topology.

***6.** If $\{(X_\alpha, \mathscr{T}_\alpha) : \alpha \in \Lambda\}$ is a collection of topological spaces, Λ is finite, and U_α is a nonempty open subset of X_α for each $\alpha \in \Lambda$, then $\mathbf{X}\{U_\alpha : \alpha \in \Lambda\}$ is open with respect to the product topology.

***7.** If $\{(X_\alpha, \mathscr{T}_\alpha) : \alpha \in \Lambda\}$ is a collection of topological spaces, Λ is infinite, and U_α is a nonempty open subset of X_α for each $\alpha \in \Lambda$, then $\mathbf{X}\{U_\alpha : \alpha \in \Lambda\}$ is an open subset of the product space.

***8.** If $\{(X_\alpha, \mathscr{T}_\alpha) : \alpha \in \Lambda\}$ is a collection of topological spaces, Λ is finite, and $A_\alpha \subseteq X_\alpha$ for each $\alpha \in \Lambda$, then $\text{Int}(\mathbf{X}\{A_\alpha : \alpha \in \Lambda\}) = \mathbf{X}\{\text{Int}(A_\alpha) : \alpha \in \Lambda\}$.

***9.** If $\{(X_\alpha, \mathscr{T}_\alpha) : \alpha \in \Lambda\}$ is a collection of topological spaces, Λ is infinite, and $A_\alpha \subseteq X_\alpha$ for each $\alpha \in \Lambda$, then $\text{Int}(\mathbf{X}\{A_\alpha : \alpha \in \Lambda\}) = \mathbf{X}\{\text{Int}(A_\alpha) : \alpha \in \Lambda\}$.

10. Let $\mathbb{R}$ have the $\mathscr{U}$ topology. The collection $\{U \times V : U, V \in \mathscr{U}\}$ is the only base for the product topology on $\mathbb{R} \times \mathbb{R}$.

11. A projection function is always a homeomorphism.

12. A projection function is always open but may not be continuous.

13. A projection function is always onto.

14. A projection function is always one-to-one.

15. If $A \subseteq \mathbb{R}$ and both $p_1(A)$ and $p_2(A)$ are open, where p_1 and p_2 are the projections of the product $(\mathbb{R}, \mathscr{U}) \times (\mathbb{R}, \mathscr{U})$ onto $(\mathbb{R}, \mathscr{U})$, then A is open in $(\mathbb{R}, \mathscr{U}) \times (\mathbb{R}, \mathscr{U})$.

***16.** A subbase for a topology is also a base for the same topology.

***17.** A base for a topology is also a subbase for the same topology.

***18.** A topology is a subbase for itself.

5

Connectedness

5.1 *Connected Spaces*

In this section a very useful topological property called connectedness is defined and developed. Roughly speaking, a topological space is connected if it cannot be divided into two "pieces." Technically the "pieces" are nonempty disjoint open subsets.

When appropriate, we are denoting a topological space $(X, \mathscr{T})$ by only the set X. We assume that a topology for X is understood.

Definition 5.1.1

A topological space X is said to be *disconnected* if there are disjoint nonempty open subsets U and V of X for which $X = U \cup V$. A topological space X is said to be *connected* if it is not disconnected.

The concept of connectedness was developed for $\mathbb{R}^n$ (with respect to the usual topology) by Camille Jordan (1838–1922) in 1892. The concept was later extended and generalized by N. J. Lennes in 1911 and by Hausdorff in 1914.

Example 5.1.2

If $X = \mathbb{R}$ with the $\mathscr{H}$ topology, then X is disconnected. To see this, note that $U = (-\infty, 0)$ and $V = [0, +\infty)$ are open sets and that $X = U \cup V$.

Example 5.1.3

If $X = \mathbb{R}$ with the $\mathscr{C}$ topology, then X is connected. There are no disjoint nonempty open subsets in this space.

Example 5.1.4

Let $X = \{a, b, c\}$ and $\mathscr{T} = \{X, \varnothing, \{a\}, \{b, c\}\}$. If $U = \{a\}$ and $V = \{b, c\}$, then U and V are disjoint open subsets whose union is X. Therefore $(X, \mathscr{T})$ is disconnected.

Example 5.1.5

Let $X = \{a, b, c\}$ and $\mathscr{T} = \{X, \varnothing, \{a\}, \{a, b\}\}$. Since there are no disjoint nonempty $\mathscr{T}$-open subsets of X, it follows that $(X, \mathscr{T})$ is connected.

Example 5.1.6

Any discrete topological space containing two or more elements is disconnected. Any indiscrete space is obviously connected.

The space $(\mathbb{R}, \mathscr{U})$ is connected. However, the proof of this fact requires the following definition and property of the real numbers.

Definition 5.1.7

Let U be a nonempty subset of $\mathbb{R}$. The set U is said to be *bounded above* if there is an element $b \in \mathbb{R}$ for which $x \leqq b$ for all $x \in U$. The element b is called an *upper bound* for U. The set U is said to be *bounded below* if there is an element $a \in \mathbb{R}$ for which $x \geq a$ for all $x \in U$. The element a is called a *lower bound* for U. The set U is said to be *bounded* if it is bounded above and bounded below. An element $m \in \mathbb{R}$ is said to be a *least upper bound* for U if m is an upper bound for U and m is less than any other upper bound for U. An element $n \in \mathbb{R}$ is said to be a *greatest lower bound* for U if n is a lower bound for U and n is greater than any other lower bound for U.

Note that if a set has a least upper bound then it is unique. To see this, assume that m and m' are least upper bounds for a subset U of $\mathbb{R}$. Since m is an upper bound and m' is a least upper bound, $m' \leqq m$. Similarly since m' is an upper bound and m is a least upper bound, $m \leqq m'$. Therefore $m = m'$. An analogous argument shows that the greatest lower bound of a set is also unique.

Property 5.1.8 (Completeness Property)

Every nonempty subset of $\mathbb{R}$ that has an upper bound also has a least upper bound. Similarly every nonempty subset of $\mathbb{R}$ that has a lower bound also has a greatest lower bound.

The next two lemmas will simplify the proof that $(\mathbb{R}, \mathscr{U})$ is connected.

LEMMA 5.1.9 *Let U be a nonempty subset of $\mathbb{R}$ that is bounded above and let m be the least upper bound for U. If I is any open interval containing m, then $I \cap U \neq \varnothing$.*

Proof Let $I = (a, b)$ be an open interval containing m. Suppose there is no element of U in the interval (a, m). Then a is an upper bound for U. This is a contradiction since $a < m$ and m is the least upper bound for U. Therefore $(a, m) \cap U \neq \emptyset$ and hence $I \cap U \neq \emptyset$. ■

LEMMA 5.1.10 *Let U be a nonempty subset of $\mathbb{R}$ that is bounded below and let n be the greatest lower bound for U. If I is any open interval containing m, then $I \cap U \neq \emptyset$.*

The proof is analogous to that of Lemma 5.1.9 and is left as an exercise.

THEOREM 5.1.11 *The topological space $(\mathbb{R}, \mathscr{U})$ is connected.*

Proof Suppose $(\mathbb{R}, \mathscr{U})$ is disconnected. Then there are disjoint nonempty $\mathscr{U}$-open subsets U and V of $\mathbb{R}$ such that $U \cup V = \mathbb{R}$. Let $a \in U$ and $b \in V$. Either $a < b$ or $a > b$. Assume $a < b$. (The proof of the case where $a > b$ is similar and is left as an exercise.)

Let $W = \{x \in U : x < b\}$. Since $a < b$, it follows that $a \in W$ and hence $W \neq \emptyset$. Because W is bounded above by b, it follows from the Completeness Property that W has a least upper bound, w. We shall show that $w \notin U$ and that $w \notin V$.

Suppose $w \in U$. Since U is open, there is an open interval (m, n) such that $w \in (m, n) \subseteq U$. Then the interval (w, n) is disjoint from W. Therefore for each $x \in (w, n)$ we have that $x \geqq b$. It follows that $w \geqq b$. However, since b is an upper bound for W and w is the least upper bound for W, we have $w \leqq b$ and hence $b = w$. Then $w \in U \cap V$ which is a contradiction since $U \cap V = \emptyset$. Thus $w \notin U$.

Suppose $w \in V$. Since V is open, there is an open interval I for which $w \in I \subseteq V$. By Lemma 5.1.9, it follows that $I \cap W \neq \emptyset$ and hence $I \cap U \neq \emptyset$. Therefore $V \cap U \neq \emptyset$ which is again a contradiction. Thus $w \notin V$.

Since $w \notin U$ and $w \notin V$, we have contradicted the statement that $(\mathbb{R}, \mathscr{U})$ is disconnected. Thus $(\mathbb{R}, \mathscr{U})$ is connected. ■

The next theorem gives another characterization of connectedness.

THEOREM 5.1.12 *A topological space is disconnected* iff *there is a nonempty proper subset of the space that is both open and closed.*

Proof $(\Rightarrow)$ Let X be a topological space. Assume X is disconnected. Then there are disjoint nonempty open subsets U and V of X such that $X = U \cup V$. Since $U = X - V$, it follows that U is closed. Since $V \neq \emptyset$, we have that U is a proper subset of X. Therefore U is a nonempty proper subset of X that is both open and closed.

$(\Leftarrow)$ Assume X has a proper nonempty subset U that is both open and closed. Then X is disconnected since U and $X - U$ are two disjoint nonempty open sets whose union is X. ■

Example 5.1.13

The subset $[0, 1)$ of $\mathbb{R}$ is both $\mathscr{H}$-open and $\mathscr{H}$-closed. Thus Theorem 5.1.12 yields a second proof that $(\mathbb{R}, \mathscr{H})$ is disconnected.

Example 5.1.14
Let $X = \{a, b, c, d\}$ and $\mathscr{T} = \{X, \varnothing, \{a, b\}, \{c, d\}\}$. Since the set $\{a, b\}$ is both $\mathscr{T}$-open and $\mathscr{T}$-closed, $(X, \mathscr{T})$ is disconnected.

Exercises 5.1

1. Let $X = \mathbb{R}$ and $\mathscr{T} = \{U \subseteq X : 1 \in U \text{ or } U = \varnothing\}$. Prove that $(X, \mathscr{T})$ is connected.
2. Let $X = \{a, b, c, d\}$ and $\mathscr{T} = \{X, \varnothing, \{a\}, \{c\}, \{a, c\}, \{b, c, d\}\}$. Prove that $(X, \mathscr{T})$ is disconnected by use of Definition 5.1.1. Then give a second proof using Theorem 5.1.12.
3. Let $X = [0, 1] \cup \{5\} \cup [8, 9]$. Show that $(X, \mathscr{U}_X)$ is disconnected.
4. Prove Lemma 5.1.10.
5. Complete the proof of Theorem 5.1.11.
6. Prove that the sets U and V in Definition 5.1.1 are closed sets.
7. Prove that if X is a set with an infinite number of elements and $\mathscr{T}$ is the finite complement topology on X, then $(X, \mathscr{T})$ is connected.
8. Prove that if $(X, \mathscr{T})$ is connected and $\mathscr{S}$ is any topology on X that is coarser than $\mathscr{T}$, then $(X, \mathscr{S})$ is connected.
9. Give an example to show that a set X can have two topologies $\mathscr{T}$ and $\mathscr{F}$ such that $\mathscr{F}$ is finer than $\mathscr{T}$, $(X, \mathscr{T})$ is connected, and $(X, \mathscr{F})$ is not connected.
10. Two nonempty subsets A and B of a topological space X are said to be *separated* provided that $\mathrm{Cl}(A) \cap B = A \cap \mathrm{Cl}(B) = \varnothing$. Prove that a space X is connected iff X is not the union of two separated sets.

5.2 *Connected Subspaces and Continuous Images of Connected Spaces*

Additional properties and applications of connectedness are presented in this section. In particular, one of the most important theorems in calculus, The Intermediate Value Theorem, is proved and some of its consequences are investigated.

Let $(X, \mathscr{T})$ be a topological space and let $A \subseteq X$. When we say that A is connected, we mean that the topological space $(A, \mathscr{T}_A)$ is connected. The following theorem characterizes the connectedness of the subset A in terms of the topology $\mathscr{T}$ on X (see Figure 5.2.1).

THEOREM 5.2.1 *Let $(X, \mathscr{T})$ be a topological space and let $A \subseteq X$. Then A is disconnected* iff *there are $\mathscr{T}$-open subsets U and V of X such that $A \subseteq U \cup V$, $A \cap U \neq \varnothing$, $A \cap V \neq \varnothing$, and $A \cap U \cap V = \varnothing$.*

Proof ($\Rightarrow$) Assume A is disconnected. Then there exist nonempty disjoint $\mathscr{T}_A$-open sets O_1 and O_2 such that $A = O_1 \cup O_2$. Since O_1 and O_2 are $\mathscr{T}_A$-open, there exist $\mathscr{T}$-open sets U and V for which $O_1 = U \cap A$ and $O_2 = V \cap A$. Then $U \cap A \neq \varnothing$ and $V \cap A \neq \varnothing$ and, since $A = O_1 \cup O_2$, obviously $A \subseteq U \cup V$. Also since O_1 and O_2 are disjoint, it follows that $A \cap U \cap V = \varnothing$.

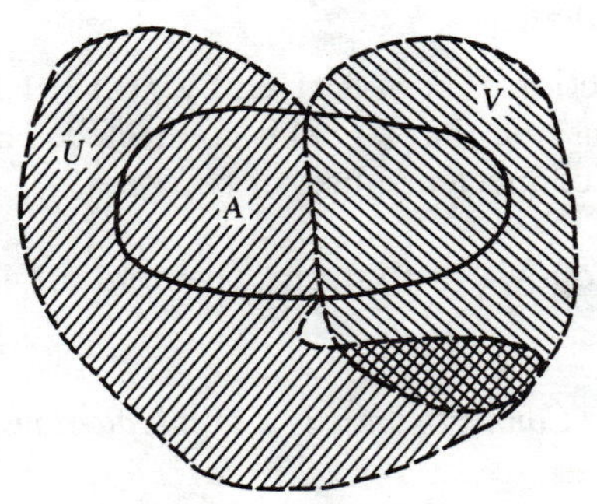

Figure 5.2.1

($\Leftarrow$) Assume there exist $\mathcal{T}$-open subsets U and V of X such that $A \subseteq U \cup V$, $A \cap U \neq \varnothing$, $A \cap V \neq \varnothing$, and $A \cap U \cap V = \varnothing$. Let $O_1 = U \cap A$ and $O_2 = V \cap A$. It is left as an exercise to show that O_1 and O_2 satisfy the conditions of Definition 5.1.1. ■

Example 5.2.2

Let $A = [0, 1] \cup \{3\} \cup (5, 6]$ be a subset of the topological space $(\mathbb{R}, \mathcal{U})$. Let $U = (-1, 5)$ and $V = (3, 7)$. Since U and V are $\mathcal{U}$-open sets for which $U \cap A \neq \varnothing, V \cap A \neq \varnothing, A \subseteq U \cup V$, and $A \cap U \cap V = \varnothing$, it follows that A is disconnected.

Example 5.2.3

The set $\mathbb{Q}$ is a disconnected subset of $(\mathbb{R}, \mathcal{U})$. To see this, note that the sets $U = (-\infty, \sqrt{2})$ and $V = (\sqrt{2}, +\infty)$ satisfy the conditions of Theorem 5.2.1.

The next theorem states that a continuous image of a connected set is connected. In other words, the quality of connectedness is preserved by continuous functions in the sense that connected sets (i.e., sets in one "piece") are not split into more than one "piece" by continuous functions.

THEOREM 5.2.4 *Let X and Y be topological spaces with $A \subseteq X$ and let $f: X \to Y$ be a continuous function. If A is connected, then $f(A)$ is connected.*

Proof Assume that A is connected. Suppose $f(A)$ is disconnected. Then there exist open subsets U and V of Y such that $f(A) \subseteq U \cup V$, $f(A) \cap U \neq \varnothing$, $f(A) \cap V \neq \varnothing$, and $f(A) \cap U \cap V = \varnothing$. Therefore $A \subseteq f^{-1}(f(A)) \subseteq f^{-1}(U \cup V) = f^{-1}(U) \cup f^{-1}(V)$. Since $f(A) \cap U \neq \varnothing$ and $f(A) \cap V \neq \varnothing$, we have that $A \cap f^{-1}(U) \neq \varnothing$ and $A \cap f^{-1}(V) \neq \varnothing$. Also $A \cap f^{-1}(U) \cap f^{-1}(V) = \varnothing$ because if $x \in A \cap f^{-1}(U) \cap f^{-1}(V)$ then $f(x) \in f(A) \cap U \cap V$ which is empty. Finally since f is continuous, $f^{-1}(U)$ and $f^{-1}(V)$ are open subsets of X. Thus by Theorem 5.2.1, A is disconnected which is a contradiction. ■

Example 5.2.5

Consider the function $f\colon \mathbb{R} \to \mathbb{R}$ given by $f(x) = 1/(1 + x^2)$. Note that f is $\mathscr{U}$-$\mathscr{U}$ continuous and that $(\mathbb{R}, \mathscr{U})$ is connected. Thus by Theorem 5.2.4, $f(\mathbb{R}) = (0, 1]$ is a connected subset of $(\mathbb{R}, \mathscr{U})$.

The next result follows easily from Theorem 5.2.4. The proof is left as an exercise.

THEOREM 5.2.6 *Connectedness is a topological property.*

Example 5.2.7

If I is any open interval, then $(I, \mathscr{U}_I)$ is connected because it is homeomorphic to $(\mathbb{R}, \mathscr{U})$.

THEOREM 5.2.8 *Let X be a topological space and let $A \subseteq X$. If A is connected and $A \subseteq B \subseteq \mathrm{Cl}(A)$, then B is connected.*

Proof Assume B is disconnected. Then there exist open subsets U and V of X such that $B \subseteq U \cup V$, $B \cap U \neq \varnothing$, $B \cap V \neq \varnothing$, and $B \cap U \cap V = \varnothing$. Since $A \subseteq B$, we have that $A \subseteq U \cup V$ and $A \cap U \cap V = \varnothing$. Since A is connected we must have $A \cap U = \varnothing$ or $A \cap V = \varnothing$. Suppose $A \cap U = \varnothing$. Since $B \cap U \neq \varnothing$ and $B \subseteq \mathrm{Cl}(A)$, it follows that U contains a point of $\mathrm{Cl}(A)$ that is not in A. That is, U contains a limit point of A. This is a contradiction because $A \cap U = \varnothing$. It is left as an exercise to show that a similar contradiction is reached if we assume that $A \cap V = \varnothing$. ■

The following corollary is an immediate consequence of Theorem 5.2.8.

COROLLARY 5.2.9 *If A is a connected subset of a topological space X, then $\mathrm{Cl}(A)$ is connected.*

Example 5.2.10

Any closed interval $[a, b]$ is a connected subset of $(\mathbb{R}, \mathscr{U})$. To see this, note that $[a, b] = \mathrm{Cl}((a, b))$ and recall that (a, b) is connected.

Although we shall not prove it, the connected subsets of $(\mathbb{R}, \mathscr{U})$ are precisely the intervals and singleton sets.

We are now able to prove one of the central theorems of calculus, The Intermediate Value Theorem. This theorem has many important consequences, some of which are presented here.

THEOREM 5.2.11 (The Intermediate Value Theorem) *Let $[a, b]$ be a closed interval and let $f\colon ([a, b], \mathscr{U}_{[a,b]}) \to (\mathbb{R}, \mathscr{U})$ be a continuous function. If y is any number between $f(a)$ and $f(b)$, then there is a number x between a and b such that $f(x) = y$.*

Proof Let y be any number between $f(a)$ and $f(b)$. Suppose that $f(x) \neq y$ for each number x between a and b. Let $U = (-\infty, y)$ and $V = (y, +\infty)$.

Obviously U and V are $\mathscr{U}$-open sets. Also since y is between $f(a)$ and $f(b)$, it follows that $f([a,b]) \cap U \neq \varnothing$ and $f([a,b]) \cap V \neq \varnothing$. Because $y \neq f(x)$ for each x between a and b, we have that $f([a,b]) \subseteq U \cup V$. Finally, since $U \cap V = \varnothing$, clearly $f([a,b]) \cap U \cap V = \varnothing$. Therefore $f([a,b])$ is disconnected. This is a contradiction since f is continuous and $[a,b]$ is connected. Hence there exists a number x between a and b for which $f(x) = y$. ■

The next two results follow easily from The Intermediate Value Theorem and are left as exercises. Recall that a *zero* of a real-valued function f is a point x such that $f(x) = 0$.

COROLLARY 5.2.12 *Let $[a,b]$ be a closed interval and let $f: ([a,b], \mathscr{U}_{[a,b]}) \to (\mathbb{R}, \mathscr{U})$ be a continuous function such that $f(a)$ and $f(b)$ have opposite signs. Then the function f has a zero between a and b.*

COROLLARY 5.2.13 *If I is any interval (bounded or unbounded) and $f: (I, \mathscr{U}_I) \to (\mathbb{R}, \mathscr{U})$ is a continuous function with no zeros in I, then either $f(x) > 0$ for all $x \in I$ or $f(x) < 0$ for all $x \in I$.*

Definition 5.2.14

A function f from a set X to itself is said to have a *fixed point* if there exists $x \in X$ such that $f(x) = x$. A topological space X is said to have the *fixed point property* provided that every continuous function from X to itself has a fixed point.

The following theorem is also a consequence of The Intermediate Value Theorem.

THEOREM 5.2.15 *A closed interval with the $\mathscr{U}$-relative topology has the fixed point property.*

Proof Let $[a,b]$ be a closed interval with the $\mathscr{U}$-relative topology and let $f: [a,b] \to [a,b]$ be a continuous function. Suppose f does not have a fixed point. Define $g: [a,b] \to \mathbb{R}$, where $\mathbb{R}$ has the usual topology, by $g(x) = f(x) - x$. Then g is clearly continuous and has no zeros in $[a,b]$. By Corollary 5.2.13 either $g(x) > 0$ for all $x \in [a,b]$ or $g(x) < 0$ for all $x \in [a,b]$. If $g(x) > 0$ for all $x \in [a,b]$, then $g(b) = f(b) - b > 0$ which contradicts $f(b) \in [a,b]$. Similarly, if $g(x) < 0$ for all $x \in [a,b]$, then $g(a) = f(a) - a < 0$ which contradicts $f(a) \in [a,b]$. Thus f must have a fixed point. ■

Geometrically the above theorem means that the graph of a continuous function from $[a,b]$ to itself must intersect the line $y = x$ (see Figure 5.2.2). We see in the next example that an open interval does not have the fixed point property. In other words, the graph of a continuous function from an open interval to itself need not intersect the line $y = x$.

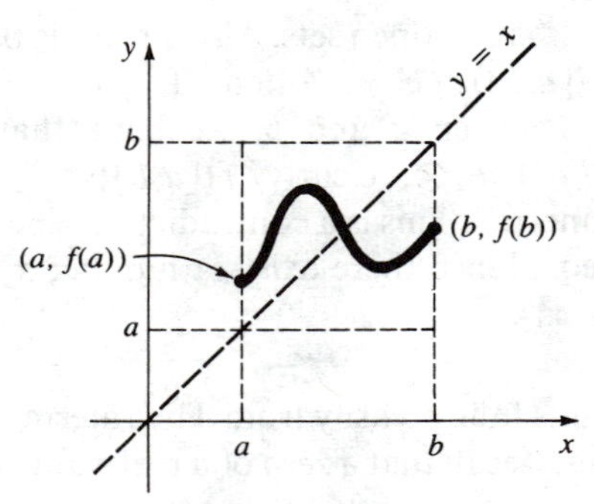

Figure 5.2.2

Example 5.2.16

Let $f: (0,1) \to (0,1)$ be given by $f(x) = x/2$. Then $f(x) = x$ implies that $2x = x$ and hence that $x = 0$. Therefore f does not have a fixed point. However, f is clearly continuous with respect to the $\mathscr{U}$-relative topology.

THEOREM 5.2.17 *The fixed point property is a topological property.*

Proof Let X and Y be homeomorphic topological spaces and let $h: X \to Y$ be a homeomorphism. Assume X has the fixed point property and let $f: Y \to Y$ be a continuous function. We must show that f has a fixed point.

Note that the composite function $h^{-1} \circ f \circ h$ is a continuous function from X to X. Since X has the fixed point property, $h^{-1} \circ f \circ h$ has a fixed point x. That is, $h^{-1}(f(h(x))) = x$ and hence $f(h(x)) = h(x)$. Thus $h(x)$ is a fixed point of f. This proves that Y has the fixed point property. ∎

Since $(\mathbb{R}, \mathscr{U})$ is homeomorphic to the space $((0,1), \mathscr{U}_{(0,1)})$ which does not have the fixed point property, it follows that $(\mathbb{R}, \mathscr{U})$ does not have the fixed point property.

Next we define a property for connected spaces that is very useful for determining if two spaces are homeomorphic.

Definition 5.2.18

A point x of a connected topological space X is said to be a *cut point* provided that $X - \{x\}$ is disconnected.

Example 5.2.19

Any point x in the space $(\mathbb{R}, \mathscr{U})$ is a cut point. Note that if $x \in \mathbb{R}$, then $\mathbb{R} - \{x\} = (-\infty, x) \cup (x, +\infty)$.

Example 5.2.20

The point 1 is a cut point of the space $([0,2], \mathscr{U}_{[0,2]})$. However, the endpoints 0 and 2 are not cut points.

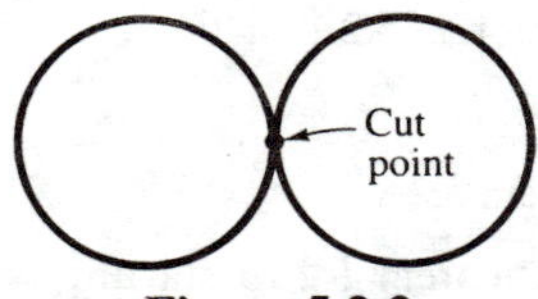

Figure 5.2.3

Example 5.2.21

If $X = \mathbb{R}$ with the $\mathscr{C}$ topology, then X has no cut points. In fact, for each $x \in X$ any two nonempty open subsets of the space $X - \{x\}$ have nonempty intersection.

A subspace of $\mathbb{R}^2$ (with the usual topology) consisting of two circles with a common point of tangency has exactly one cut point (see Figure 5.2.3).

THEOREM 5.2.22 *Let X and Y be topological spaces and let $f: X \to Y$ be a homeomorphism. If x is a cut point of X, then $f(x)$ is a cut point of Y.*

The proof is left as an exercise.

Because of this theorem the concept of a cut point can be used to determine if two topological spaces are homeomorphic.

Example 5.2.23

The line segment and the circle in Figure 3.4.4 (page 63) are not homeomorphic. The circle has no cut points, but every point of the line segment except the endpoints is a cut point.

Example 5.2.24

If $\mathbb{R}$ and $\mathbb{R}^2$ have their usual topologies, then $\mathbb{R}$ is not homeomorphic to $\mathbb{R}^2$. Note that every point of $\mathbb{R}$ is a cut point, but $\mathbb{R}^2$ has no cut points.

Exercises 5.2

1. Prove that the set $A = \{1, 2\}$ is a disconnected subset of $(\mathbb{R}, \mathscr{U})$.
2. Show that $B = (-\infty, 0] \cup \{3\} \cup [6, 8]$ is a disconnected subset of $(\mathbb{R}, \mathscr{U})$.
3. Let X be a topological space with $x \in X$. Show that $\{x\}$ is connected.
4. Show that the set of irrational numbers is a disconnected subset of $(\mathbb{R}, \mathscr{U})$.
5. Let $f: \mathbb{R} \to \mathbb{R}$ be given by $f(x) = 4/(2 + x^2)$. Assume that f is $\mathscr{U}$-$\mathscr{U}$ continuous. Use a method similar to that of Example 5.2.5 to show that $(0, 2]$ is a connected subset of $(\mathbb{R}, \mathscr{U})$.
6. Let X be a connected topological space with $a, b \in X$ and let $f: X \to \mathbb{R}$ be a continuous function (where $\mathbb{R}$ has the usual topology). Prove that if y is any number between $f(a)$ and $f(b)$, then there is an element $x \in X$ such that $f(x) = y$. (This is a slightly stronger version of The Intermediate Value Theorem.)

7. Complete the proof of Theorem 5.2.1.

8. Prove Theorem 5.2.6.

9. Finish the proof of Theorem 5.2.8.

10. Give a direct proof of Theorem 5.2.15 starting with The Intermediate Value Theorem.

11. Prove Corollary 5.2.12.

12. Prove Corollary 5.2.13.

13. Determine if The Intermediate Value Theorem holds if the $\mathscr{U}$-relative topology on $[a, b]$ is replaced by the $\mathscr{H}$-relative topology.

14. Assume the interval $(1, 2)$ has the $\mathscr{U}$-relative topology. Give an example of a continuous function $f\colon (1, 2) \to (1, 2)$ that does not have a fixed point.

15. Give an example of a function $f\colon (\mathbb{R}, \mathscr{U}) \to (\mathbb{R}, \mathscr{U})$ that is continuous and does not have a fixed point.

16. Determine if the space $([0, 1], \mathscr{H}_{[0,1]})$ has the the fixed point property.

17. Give an example to show that the union of two connected sets is not necessarily connected.

18. Give an example to show that the intersection of two connected sets is not necessarily connected.

19. Give an example to show that the union of two disconnected sets can be connected.

20. Let X be any set with the indiscrete topology. Show that X has no cut points.

21. Let $X = \{a, b, c\}$ and $\mathscr{T} = \{X, \varnothing, \{a\}, \{b\}, \{a, b\}\}$. Find all cut points of $(X, \mathscr{T})$.

22. Prove Theorem 5.2.22.

23. In Theorem 5.2.22 determine if the requirement that f be a homeomorphism can be replaced by the condition that f be a continuous function.

24. Use the concept of a cut point to show that no two of the spaces in Figure 3.4.5 are homeomorphic.

5.3 *Finite Products of Connected Spaces*

We shall prove that the product of two connected topological spaces (and hence any finite number of spaces) is connected. Since the proofs are somewhat complicated, several lemmas will be proved first.

Throughout this section and the following section we shall assume that the subset $\{0, 1\}$ of $\mathbb{R}$ has the discrete topology, which is the same as the relative topology with respect to $\mathscr{U}$. Note that the only nonempty connected subsets of $\{0, 1\}$ are $\{0\}$ and $\{1\}$. This fact is used several times in the following proofs.

LEMMA 5.3.1 *A topological space X is connected* iff *the only continuous functions from X to $\{0, 1\}$ are the constant functions.*

Proof ($\Rightarrow$) Assume X is connected. Let $f: X \to \{0, 1\}$ be a continuous function. Suppose f is nonconstant. Let $U = f^{-1}(0)$ and $V = f^{-1}(1)$. Since f is continuous, U and V are open. Because f is nonconstant, U and V are nonempty. Also since f is a function, it follows that $X = U \cup V$ and that U and V are disjoint. Therefore X is disconnected which is a contradiction. Hence the only continuous functions from X to $\{0, 1\}$ are the constant functions.

($\Leftarrow$) Assume that the only continuous functions from X to $\{0, 1\}$ are the constant functions. Suppose X is disconnected. Then there exist disjoint nonempty open subsets U and V of X for which $X = U \cup V$. Define the function $f: X \to \{0, 1\}$ by

$$f(x) = \begin{cases} 0 & \text{if } x \in U \\ 1 & \text{if } x \in V \end{cases}$$

Since U and V are disjoint, f is a well-defined function. It is left as an exercise to show that f is continuous. Note that since U and V are nonempty, it follows that f is nonconstant and hence we have reached a contradiction. Therefore X is connected. ■

LEMMA 5.3.2 *Let X and Y be topological spaces and let $y \in Y$. If X is connected, then the set $X \times \{y\}$ is a connected subset of the product space $X \times Y$.*

Proof Define the function $f: X \to X \times Y$ by $f(x) = (x, y)$. By Exercise 5, f is continuous. Since X is connected it follows from Theorem 5.2.4 that $X \times \{y\} = f(X)$ is connected. ■

Figure 5.3.1 gives a geometric representation of the set $X \times \{y\}$. Note that $X \times \{y\}$ is actually homeomorphic to X. In fact, if the codomain of the function f in the proof of Lemma 5.3.2 is restricted to $X \times \{y\}$, then f is a homeomorphism (see Exercise 9 in Section 4.2).

THEOREM 5.3.3 *If X and Y are nonempty topological spaces, then the product space $X \times Y$ is connected* iff *X and Y are connected.*

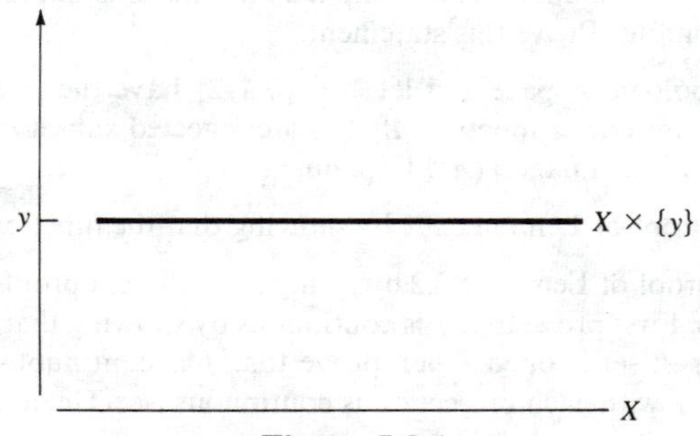

Figure 5.3.1

Proof ($\Leftarrow$) Assume X and Y are connected. Let $f: X \times Y \to \{0, 1\}$ be a continuous function. Suppose f is nonconstant. Then there are points (x_0, y_0) and (x_1, y_1) in $X \times Y$ such that $f(x_0, y_0) = 0$ and $f(x_1, y_1) = 1$. Consider the point (x_0, y_1) in $X \times Y$. Either $f(x_0, y_1) = 0$ or $f(x_0, y_1) = 1$. Suppose $f(x_0, y_1) = 0$. By Lemma 5.3.2 the set $X \times \{y_1\}$ is connected. Since f is continuous, $f(X \times \{y_1\})$ is a connected subset of $\{0, 1\}$. Since $\{0\}$ and $\{1\}$ are the only nonempty connected subsets of $\{0, 1\}$, it follows that $f(X \times \{y_1\}) = \{0\}$ or $f(X \times \{y_1\}) = \{1\}$. However, this is not possible because both (x_1, y_1) and (x_0, y_1) are in $X \times \{y_1\}$, and $f(x_1, y_1) = 1$ and $f(x_0, y_1) = 0$. Suppose $f(x_0, y_1) = 1$. It follows from Lemma 5.3.2 that $\{x_0\} \times Y$ is connected. Since f is continuous, we have that $f(\{x_0\} \times Y)$ is a connected subset of $\{0, 1\}$. Therefore $f(\{x_0\} \times Y) = \{0\}$ or $f(\{x_0\} \times Y) = 1$. This is also a contradiction because both (x_0, y_0) and (x_0, y_1) are in $\{x_0\} \times Y$, and $f(x_0, y_0) = 0$ and $f(x_0, y_1) = 1$. Therefore f must be constant and hence by Lemma 5.3.1 $X \times Y$ is connected.

($\Rightarrow$) Assume $X \times Y$ is connected. Recall that the projections $p_1: X \times Y \to X$ and $p_2: X \times Y \to Y$ are continuous functions. It follows that $X = p_1(X \times Y)$ and $Y = p_2(X \times Y)$ are connected. ■

A simple induction argument proves the following theorem.

THEOREM 5.3.4 *If $X_1, X_2, \ldots, X_n$ are nonempty topological spaces, then the product space $X_1 \times X_2 \times \cdots \times X_n$ is connected* iff *X_i is connected for each $i \in \{1, 2, \ldots, n\}$.*

Exercises 5.3

1. Determine which of the following product spaces are connected. Explain your answers.

(a) $X \times Y$ where $X = \mathbb{R}$ has the $\mathscr{U}$ topology and $Y = \mathbb{R}$ has the $\mathscr{H}$ topology.
(b) $X \times X$ where $X = \mathbb{R}$ has the $\mathscr{C}$ topology.
(c) $X \times Y$ where $X = \{a, b, c\}$ has the topology $\mathscr{T} = \{X, \varnothing, \{a\}, \{a, b\}\}$, and $Y = \{d, e, f\}$ has the topology $\mathscr{S} = \{Y, \varnothing, \{d\}, \{e, f\}\}$.

2. In the statement of Lemma 5.3.1 it is implied that the constant functions from X to $\{0, 1\}$ are continuous. Prove this statement.

3. Let X be a topological space and let $Y = \{0, 1, 2\}$ have the $\mathscr{D}$ topology. Assume $f: X \to Y$ is a continuous function. If A is a connected subset of X, what are the possible values of the image $f(A)$? Explain.

4. Complete the proof of Lemma 5.3.1 by showing that the function f is continuous.

5. Complete the proof of Lemma 5.3.2 by giving two different proofs that the function f is continuous. First prove that f is continuous by showing that the inverse image of any basic open set is open. Then prove that f is continuous by showing that composition of f with each projection is continuous (see Theorem 4.2.13).

6. Let $\mathbb{R}$ have the $\mathscr{U}$ topology. By Lemma 5.3.2 the set $\mathbb{R} \times \{1\}$ is a connected subset of $\mathbb{R} \times \mathbb{R}$. However $\mathbb{R} \times \{0, 1\}$ is disconnected. Give a direct proof of this fact using the definition of connectedness.

5.4* *Infinite Products of Connected Spaces*

Our next goal is to prove that the product of an arbitrary collection of connected spaces is connected. Because of the nature of the product of an infinite number of topological spaces, the proofs in this section are very involved. We begin with two lemmas that will simplify the process. The following lemma is analogous to Lemma 5.3.2.

LEMMA 5.4.1 *Let $\{X_\alpha : \alpha \in \Lambda\}$ be a collection of connected topological spaces. Let F be a finite subset of Λ. For each $\alpha \in \Lambda - F$ let $y_\alpha \in X_\alpha$. If $A_\alpha = X_\alpha$ for each $\alpha \in F$ and $A_\alpha = \{y_\alpha\}$ for each $\alpha \in \Lambda - F$, then the set $\mathbf{X}\{A_\alpha : \alpha \in \Lambda\}$ is connected.*

Proof Let $A = \mathbf{X}\{A_\alpha : \alpha \in F\}$ and $X = \mathbf{X}\{X_\alpha : \alpha \in \Lambda\}$. Define the function $f: A \to X$ by $f(x)_\alpha = x_\alpha$ if $\alpha \in F$ and $f(x)_\alpha = y_\alpha$ if $\alpha \in \Lambda - F$. By Exercise 2, f is continuous. It follows from Theorem 5.3.4 that A is connected. Therefore $\mathbf{X}\{A_\alpha : \alpha \in \Lambda\} = f(A)$ is connected. ■

The set $\mathbf{X}\{A_\alpha : \alpha \in \Lambda\}$ is actually a "homeomorphic copy" of A in the same way that $X \times \{y\}$ is a "homeomorphic copy" of X in Lemma 5.3.2.

The next lemma uses Lemma 5.4.1 to show that if two points in a product of connected spaces differ in only a finite number of coordinates, then the points have the same image under a continuous function into the set $\{0, 1\}$. (We continue to assume that the set $\{0, 1\}$ has the discrete topology.)

LEMMA 5.4.2 *Let $\{X_\alpha : \alpha \in \Lambda\}$ be a collection of connected topological spaces and let x and y be points in the product space $X = \mathbf{X}\{X_\alpha : \alpha \in \Lambda\}$ such that $x_\alpha = y_\alpha$ except for $\alpha \in F$ where F is a finite subset of Λ. If $f: X \to \{0, 1\}$ is a continuous function, then $f(x) = f(y)$.*

Proof Let $A_\alpha = X_\alpha$ for each $\alpha \in F$ and $A_\alpha = \{x_\alpha\}$ for each $\alpha \in \Lambda - F$. By Lemma 5.4.1 the set $B = \mathbf{X}\{A_\alpha : \alpha \in \Lambda\}$ is connected. Since f is continuous, $f(B)$ is a connected subset of $\{0, 1\}$. Therefore $f(B) = \{0\}$ or $f(B) = \{1\}$. Since x and y are points in B, it follows that $f(x) = f(y)$. ■

We are now ready to prove that the product of an arbitrary collection of connected topological spaces is connected.

THEOREM 5.4.3 *If $\{X_\alpha : \alpha \in \Lambda\}$ is a collection of nonempty topological spaces, then the product space $X = \mathbf{X}\{X_\alpha : \alpha \in \Lambda\}$ is connected* iff *X_α is connected for each $\alpha \in \Lambda$.*

Proof ($\Leftarrow$): Assume X_α is connected for each $\alpha \in \Lambda$. Let $f: X \to \{0, 1\}$ be a continuous function. We shall show that f must be constant. Let $x \in X$. Assume that $f(x) = 0$. Let y be any other point of X. We shall show that $f(y) = 0$ also. Note that since $\{0, 1\}$ has the discrete topology, $\{0\}$ is an open subset of $\{0, 1\}$. Since f is continuous, $f^{-1}(\{0\})$ is an open subset of the product space X. Therefore there is a basic open set of the form $U = \mathbf{X}\{U_\alpha : \alpha \in \Lambda\}$ where U_α is an open subset of X_α for each $\alpha \in \Lambda$ and $U_\alpha = X_\alpha$ except for $\alpha \in F$ where F is a finite subset of Λ and for which $x \in U \subseteq f^{-1}(\{0\})$. Note that for $\alpha \in \Lambda - F$, the αth coordinates of points in U are unrestricted. Therefore there is an element z in U such that for each $\alpha \in \Lambda - F$, $z_\alpha = y_\alpha$. Since $z_\alpha = y_\alpha$ except for a finite number of α, it follows from Lemma 5.4.2 that $f(z) = f(y)$. Because $z \in f^{-1}(\{0\})$, we have that $f(z) = 0$. Therefore $f(y) = 0$. Hence f is constant and by Lemma 5.3.1, X is connected.

The proof that if the product space X is connected, then X_α is connected for each $\alpha \in \Lambda$ is left as an exercise. ■

Exercises 5.4

1. Determine which of the following product spaces are connected. Explain your answers.
   (a) $\mathbf{X}\{X_i : i \in \Lambda\}$, where $\Lambda = \mathbb{Z}^+$ and for each $i \in \Lambda$, $X_i = \mathbb{R}$ has the $\mathscr{H}$ topology.
   (b) $\mathbf{X}\{Y_\alpha : \alpha \in \Lambda\}$, where $\Lambda = \mathbb{R}^+$ and for each $\alpha \in \Lambda$, $Y_\alpha = \mathbb{R}$ has the $\mathscr{C}$ topology for $\alpha \leqq 1$ and the discrete topology for $\alpha > 1$.
   (c) $\mathbf{X}\{Z_i : i \in \Lambda\}$, where $\Lambda = \mathbb{Z}^+$ and for each $i \in \Lambda$, $Z_i = \mathbb{R}$ has the topology $\mathscr{T}_i = \{U \subseteq Z_i : i \in U \text{ or } U = \varnothing\}$.
   (d) $\mathbf{X}\{W_\alpha : \alpha \in \Lambda\}$, $\Lambda = [0, 1]$ and for each $\alpha \in \Lambda$, $W_\alpha = [0, 1] - \{\alpha\}$ with the $\mathscr{U}$-relative topology.
2. Show that the function f in the proof of Lemma 5.4.1 is continuous by showing that the composition of f with each projection is continuous (see Theorem 4.3.14).
3. Complete the proof of Theorem 5.4.3.
4. Prove that the product space $\mathbf{X}\{X_\alpha : \alpha \in \Lambda\}$ is disconnected if there exists $\alpha_0 \in \Lambda$ for which X_{α_0} is disconnected.

Review Exercises 5

Mark each of the following statements true or false. Briefly explain each true statement and find a counterexample for each false statement.

1. Any discrete topological space is disconnected.
2. Any discrete topological space with more than one element is disconnected.
3. Any indiscrete topological space is connected.
4. Both of the spaces $(\mathbb{R}, \mathscr{U})$ and $(\mathbb{R}, \mathscr{H})$ are connected.
5. Any subset of the space $(\mathbb{R}, \mathscr{U})$ is connected.

6. If $X = \mathbb{R}$ has the usual topology, Y is a topological space, and $f: X \to Y$ is a continuous function, then $f(\{1, 2\})$ is a connected subset of Y.

7. If $X = \mathbb{R}$ has the $\mathscr{C}$ topology, Y is a topological space, and $f: X \to Y$ is a continuous function, then $f(\{1, 2\})$ is a connected subset of Y.

8. If $X = \mathbb{R}$ has the usual topology, Y is a topological space, and $f: X \to Y$ is a one-to-one function, then $f([0, 1])$ is a connected subset of Y.

9. If $X = \mathbb{R}$ has the usual topology, Y is a topological space, and $f: X \to Y$ is a continuous function, then $f([0, 1])$ is a connected subset of Y.

10. If X and Y are topological spaces with $A \subseteq X$ and $f: X \to Y$ is a continuous function, then A is connected if $f(A)$ is connected.

11. If X and Y are topological spaces with $A \subseteq X$ and $f: X \to Y$ is a continuous function, then $f(A)$ is connected if A is connected.

12. If X and Y are topological spaces with $B \subseteq Y$ and $f: X \to Y$ is a continuous function, then $f^{-1}(B)$ is connected if B is connected.

13. Every continuous function from $([0, 1], \mathscr{U}_{[0,1]})$ to itself has a fixed point.

14. Every function from $([0, 1], \mathscr{U}_{[0,1]})$ to itself has a fixed point.

15. Every continuous function from $((0, 1], \mathscr{U}_{(0,1]})$ to itself has a fixed point.

16. Every continuous function from $([0, 1], \mathscr{U}_{[0,1]})$ to $([0, 2], \mathscr{U}_{[0,2]})$ has a fixed point.

17. Any nonempty topological space has at least one cut point.

18. A topological space can have more than one cut point.

19. If X and Y are topological spaces and X is connected, then $X \times Y$ is connected.

20. If X and Y are nonempty topological spaces and $X \times Y$ is connected, then X is connected.

21. If $\mathbb{R}$ has the usual topology and f is a one-to-one continuous function from $\mathbb{R}$ onto $\mathbb{R}$, then f is a homeomorphism.

6

Compactness

6.1 *Compact Spaces*

Subsets of the space $(\mathbb{R}, \mathscr{U})$ that are both closed and bounded have many useful properties. For example, any infinite subset of a closed and bounded subset A has a limit point in A. Also any continuous real-valued function with a closed and bounded domain has a closed and bounded range. In this section we generalize the notion of a set being closed and bounded. A topological property called compactness is developed for a general topological space. For subsets of the space $(\mathbb{R}, \mathscr{U})$ compactness turns out to be equivalent to the property of being closed and bounded. However, these concepts cannot be used to define compactness in a general topological space.

Definition 6.1.1

Let X be a topological space and let $A \subseteq X$. A collection $\mathfrak{C} = \{U_\alpha : \alpha \in \Lambda\}$ of subsets of X is said to be a *cover* (or *covering*) of A provided that $A \subseteq \bigcup\{U_\alpha : \alpha \in \Lambda\}$. If U_α is an open subset of X for each $\alpha \in \Lambda$, then $\mathfrak{C}$ is called an *open cover* of A. If $\mathfrak{C}$ and $\mathfrak{D}$ are covers of A, then $\mathfrak{D}$ is said to be a *subcover* of $\mathfrak{C}$ if $\mathfrak{D} \subseteq \mathfrak{C}$.

Example 6.1.2

Let $X = \mathbb{R}$ have the $\mathscr{U}$ topology. Each of the collections

$$\mathfrak{C} = \{(-x, x) : x \in \mathbb{R}^+\}$$

and $\mathfrak{D} = \{(-n, n) : n \in \mathbb{Z}^+\}$ is an open cover of X and $\mathfrak{D}$ is a subcover of $\mathfrak{C}$.

Example 6.1.3

Let $X = \mathbb{R}$ have the $\mathscr{H}$ topology. The collection $\mathfrak{C} = \{[n, n+1) : n \in \mathbb{Z}\}$ is an open cover of X. Note that the cover $\mathfrak{C}$ does not have a proper subcover.

Example 6.1.4

Let $X = \mathbb{R}$ have the $\mathscr{C}$ topology. Then $\mathfrak{D} = \{(-n, +\infty) : n \in \mathbb{Z}^+\}$ is an open cover of X.

Example 6.1.5

Let $X = \mathbb{R}$ have the topology $\mathscr{T} = \{U \subseteq \mathbb{R} : U = \varnothing \text{ or } 1 \in U\}$. The collection $\mathfrak{C} = \{\{1, x\} : x \in \mathbb{R} - \{1\}\}$ is an open cover of X.

The following definition of compactness was developed by P. S. Alexandroff and Paul Urysohn (1898–1924) in 1923.

Definition 6.1.6

A topological space X is said to be *compact* if every open cover of X has a finite subcover.

The above definition is sometimes difficult to use because of the complexities involved in dealing with covers and subcovers. However, the concept of compactness, as given in Definition 6.1.6, has turned out to be one of the most important and most useful of all topological properties. We shall see that compact topological spaces have many highly desirable properties.

Example 6.1.7

The space $(\mathbb{R}, \mathscr{U})$ is not compact because the open cover $\mathfrak{C}$ given in Example 6.1.2 does not have a finite subcover.

Example 6.1.8

The space $(\mathbb{R}, \mathscr{H})$ is not compact. The cover given in Example 6.1.3 does not have a finite subcover.

Example 6.1.9

Let $X = \mathbb{R}$ with the the topology $\mathscr{T} = \{X, \varnothing, (-\infty, 0), [0, +\infty)\}$. Obviously $(X, \mathscr{T})$ is compact because $\mathscr{T}$ is finite.

Example 6.1.10

Let $X = \mathbb{R}$ with the topology $\mathscr{T} = \{U \subseteq X : U = X \text{ or } 1 \notin U\}$. The space X is compact. To see this, let $\mathfrak{C}$ be any open cover of X. Then there exists $U \in \mathfrak{C}$ such that $1 \in U$. From the definition of $\mathscr{T}$ we must have $U = X$. Therefore $\{U\}$ is a finite subcover of $\mathfrak{C}$.

When we say that a subset A of a topological space $(X, \mathscr{T})$ is compact, we mean that the space $(A, \mathscr{T}_A)$ is compact. The next theorem characterizes the compactness of a subset A in terms of the open subsets of X. The proof is straightforward and is left as an exercise.

THEOREM 6.1.11 *Let $(X, \mathscr{T})$ be a topological space and let $A \subseteq X$. The set A is compact* iff *every cover of A consisting of $\mathscr{T}$-open subsets of X has a finite subcover.*

When proving that a subset of a topological space is compact, we shall use Theorem 6.1.11. In other words, if A is a subset of a space X, any open cover of A will consist of open subsets of X.

THEOREM 6.1.12 *If $\mathbb{R}$ has the usual topology, then any closed interval is a compact subset.*

Proof Let $[a, b]$ be a closed interval and $\mathfrak{C} = \{U_\alpha \subseteq \mathbb{R} : \alpha \in \Lambda\}$ be any open cover of $[a, b]$. Let A be the set of all $x \in (a, b]$ for which the interval $[a, x]$ is contained in the union of a finite number of sets in $\mathfrak{C}$. Let $\beta \in \Lambda$ such that $a \in U_\beta$. Since U_β is open, there exists an open interval I for which $a \in I \subseteq U_\beta$. Therefore there exists a number y such that $[a, y] \subseteq I \subseteq U_\beta$. Thus $y \in A$ and hence $A \neq \varnothing$. Since A is bounded above (by b), A has a least upper bound m. We shall show that $m = b$ and then that $b \in A$ (see Figure 6.1.1).

Since b is an upper bound for A and m is the least upper bound for A, clearly $m \leqq b$. Suppose $m < b$. Note that $m \in [a, b]$. Let $\gamma \in \Lambda$ such that $m \in U_\gamma$. There exists an open interval J such that $m \in J \subseteq U_\gamma$. By Lemma 5.1.9 there exists $x \in J \cap A$. Then $[a, x]$ is contained in the union of a finite number of sets in $\mathfrak{C}$. Let $z \in J \cap [a, b]$ such that $z > x$ and $z > m$. Then $[x, z] \subseteq J \subseteq U_\gamma$. Since $[a, x]$ is contained in the union of a finite

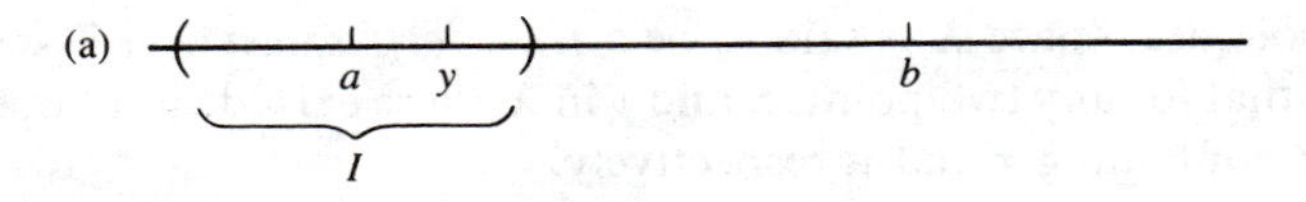

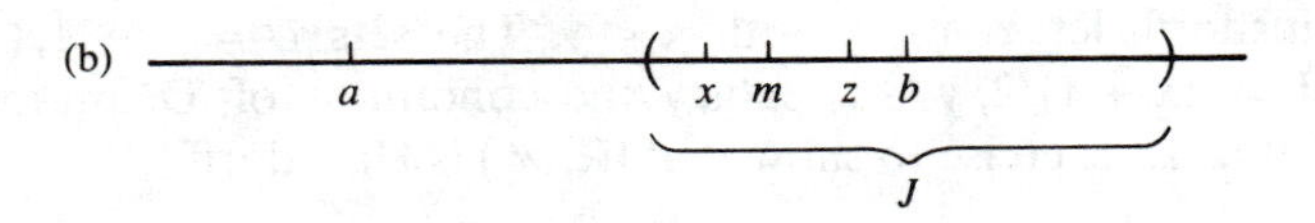

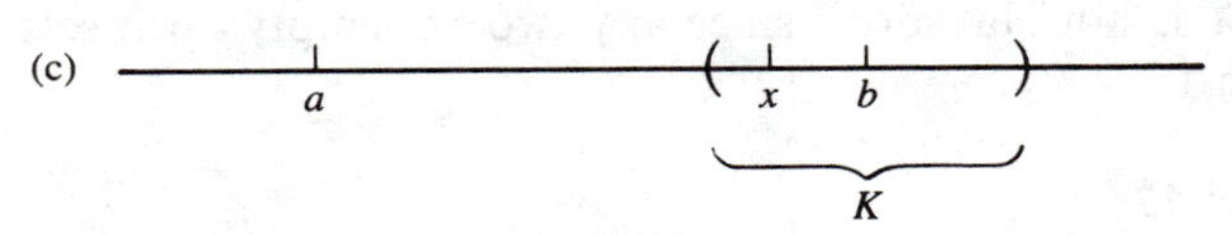

Figure 6.1.1

number of sets in $\mathfrak{C}$ and $[x, z] \subseteq U_\gamma$, we have that $[a, z]$ is contained in the union of a finite number of sets in $\mathfrak{C}$. This implies that $z \in A$ which is a contradiction because $z > m$. Therefore $m = b$.

Let $\alpha \in \Lambda$ such that $b \in U_\alpha$ and let K be an open interval for which $b \in K \subseteq U_\alpha$. Since b is the least upper bound of A, there exists $x \in K \cap A$. Then $[a, x]$ is contained in the union of a finite number of sets in $\mathfrak{C}$. Since $[x, b] \subseteq K \subseteq U_\alpha$, it follows that $[a, b]$ is contained in the union of a finite number of sets in $\mathfrak{C}$. Therefore $b \in A$ and $[a, b]$ is compact. ■

Subsets of compact spaces are not necessarily compact. However, in the next theorem we see that certain subsets of compact spaces are indeed compact.

THEOREM 6.1.13 *A closed subset of a compact topological space is compact.*

Proof Let X be a compact topological space and let A be a closed subset of X. Let $\mathfrak{C}$ be an open cover of A. Let $\mathfrak{D} = \mathfrak{C} \cup \{X - A\}$. Since $\mathfrak{C}$ is a cover of A, obviously $\mathfrak{D}$ is a cover of X. Because $\mathfrak{C}$ is an open cover and $X - A$ is open, $\mathfrak{D}$ is an open cover of X. Since X is compact, there exists a finite subcover $\mathfrak{D}'$ of $\mathfrak{D}$. Then $\mathfrak{D}' - \{X - A\}$ is a finite subcover of $\mathfrak{C}$. Therefore A is compact. ■

The property given in the following definition is an example of a separation property. These properties will be developed in detail in Chapter 7. This particular property is useful in the development of compactness.

Definition 6.1.14

A topological space X is said to be a *Hausdorff space* or a *T_2-space* provided that for any two points x and y in X there exist disjoint open sets U and V containing x and y, respectively.

Example 6.1.15

Both of the spaces $(\mathbb{R}, \mathscr{U})$ and $(\mathbb{R}, \mathscr{H})$ are Hausdorff. To see that $(\mathbb{R}, \mathscr{U})$ Hausdorff, let x, $y \in \mathbb{R}$ with $x < y$. The sets $U = (x - 1, (x + y)/2)$ and $V = ((x + y)/2, y + 1)$ satisfy the conditions of Definition 6.1.14. It is left as an exercise to show that $(\mathbb{R}, \mathscr{H})$ is Hausdorff.

Example 6.1.16

Let $X = \mathbb{R}$ with the topology $\mathscr{T} = \{U \subseteq X : U = \varnothing \text{ or } 1 \in U\}$. The space X is not Hausdorff since any two nonempty open sets must each contain 1.

Example 6.1.17

Let $X = \mathbb{R}$ with the $\mathscr{C}$ topology. The space X is not Hausdorff since any two nonempty open sets must intersect.

Our next goal is to prove that a compact subset of a Hausdorff space is closed. The following lemma will simplify the proof of this result. The lemma will also be used in the next chapter.

LEMMA 6.1.18 *If A is a compact subset of a Hausdorff space X and $x \in X - A$, then there exist disjoint open sets U and V containing A and x, respectively.*

Proof Since X is Hausdorff, for each $a \in A$ there exist disjoint open sets U_a and V_a containing a and x, respectively. The collection $\{U_a : a \in A\}$ is an open cover of A. Because A is compact, this collection has a finite subcover, $\{U_{a_i} : i = 1, 2, \ldots, n\}$. Let $U = \bigcup\{U_{a_i} : i = 1, 2, \ldots, n\}$ and let $V = \bigcap\{V_{a_i} : i = 1, 2, \ldots, n\}$. Note that $A \subseteq U$, $x \in V$, and both U and V are open sets. Also since U_{a_i} and V_{a_i} are disjoint for each $i \in \{1, 2, \ldots, n\}$, the sets U and V are disjoint. ■

THEOREM 6.1.19 *A compact subset of a Hausdorff space is closed.*

Proof Let X be a Hausdorff space and let A be a compact subset of X. Let $x \in X - A$. By Lemma 6.1.18 there exist disjoint open sets U_x and V_x containing x and A, respectively. Therefore $x \in U_x \subseteq X - V_x \subseteq X - A$. It follows that $X - A$ is open and thus A is closed. ■

The following theorem characterizes compact subsets of $(\mathbb{R}, \mathscr{U})$.

THEOREM 6.1.20 (The Heine–Borel Theorem) *Let $\mathbb{R}$ have the usual topology. A subset A of $\mathbb{R}$ is compact* iff *A is closed and bounded.*

Proof ($\Leftarrow$) Let $A \subseteq \mathbb{R}$. Assume A is closed and bounded. Because A is bounded, there is a closed interval $[a, b]$ that contains A. Since A is a closed subset of $\mathbb{R}$, A is obviously a closed subset of $[a, b]$ (see Exercise 14). By Theorem 6.1.12 $[a, b]$ is compact. Since A is a closed subset of $[a, b]$, Theorem 6.1.13 implies that A is a compact subset of $[a, b]$ and therefore also a compact subset of $\mathbb{R}$ (see Exercise 15).

($\Rightarrow$) Assume A is compact. Since $\mathbb{R}$ is Hausdorff, it follows from Theorem 6.1.19 that A is closed. To see that A is bounded, note that the collection $\mathfrak{C} = \{(-x, x) : x \in \mathbb{R}^+\}$ is an open cover of $\mathbb{R}$ and hence also an open cover of A. Because A is compact, $\mathfrak{C}$ has a finite subcover $\mathfrak{C}' = \{(-x_i, x_i) : i = 1, 2, \ldots, n\}$. Let y be the largest element of the set $\{x_i : i = 1, 2, \ldots, n\}$. Obviously $A \subseteq [-y, y]$ and hence A is bounded. ■

Accordingly, the noncompactness of a set A in $(\mathbb{R}, \mathscr{U})$ is always exhibited by either A not being bounded or A not containing one of its limit points. If A is unbounded, then the open cover $\{(p - 1, p + 1) : p \in A\}$ has no finite subcover and, if A does not contain a limit point p, then the open cover $\left\{\left(-\infty, p - \frac{1}{n}\right) \cup \left(p + \frac{1}{n}, +\infty\right) : n \in \mathbb{Z}^+\right\}$ has no finite subcover.

Eduard Heine (1821–1881) and Emile Borel (1871–1956), for whom the above theorem is named, did not collaborate in the usual way. A. M. Schoenflies (1858–1923) gave the theorem its name after observing that the result was related to work done separately by Heine and Borel.

THEOREM 6.1.21 *If X is a compact topological space and A is an infinite subset of X, then A has at least one limit point in X.*

Proof Suppose A has no limit points in X. Then for each $x \in X$, there is an open set U_x containing x such that $U_x \cap A \subseteq \{x\}$. That is, U_x does not contain any point of A except possibly x. The collection $\{U_x : x \in X\}$ is an open cover of X. Since X is compact, the collection has a finite subcover $\{U_{x_i} : i = 1, 2, \ldots, n\}$. Then $A \subseteq \bigcup\{U_{x_i} : i = 1, 2, \ldots, n\}$. This is a contradiction because A is infinite, but each set U_{x_i} contains at most one element of A. Therefore A must have at least one limit point in X. ■

The proof of the last theorem of this section is left as an exercise.

THEOREM 6.1.22 (The Bolzano–Weierstrass Theorem) *Every bounded infinite subset of $(\mathbb{R}, \mathscr{U})$ has at least one limit point.*

Exercises 6.1

1. Which of the following collections are open covers for $\mathbb{R}$ with respect to the usual topology? If a collection is not an open cover, explain why it is not.
   (a) $\{(-2n, 2n) : n \in \mathbb{Z}^+\}$ **(b)** $\{\{x\} : x \in \mathbb{R}\}$
   (c) $\{(n, n+1) : n \in \mathbb{Z}\}$ **(d)** $\{[n, n+1) : n \in \mathbb{Z}\}$
   (e) $\{(x, +\infty) : x \in \mathbb{R}\}$
2. Let $\mathbb{R}$ have the $\mathscr{H}$ topology. Let $\mathfrak{C} = \{[a, b) \subseteq \mathbb{R} : b - a \leqq 4\}$. Show that $\mathfrak{C}$ is an open cover for $\mathbb{R}$. Determine which of the following collections are subcovers of $\mathfrak{C}$. If a collection is not a subcover of $\mathfrak{C}$, explain why it is not.
   (a) $\{[n-1, n+1) : n \in \mathbb{Z}\}$ **(b)** $\{[n, n+1) : n \in \mathbb{Z}^+\}$
   (c) $\{[x-1, x+4) : x \in \mathbb{R}\}$ **(d)** $\{[x, y) : y - x < 1\}$
   (e) $\{(-x, x) : x \in \mathbb{R}^+\}$
3. Prove that any finite topological space is compact. Give an example of an infinite space that is compact.
4. Prove that any topological space with a finite topology is compact. Give an example of a compact space that has an infinite topology.
5. Let $X = \mathbb{R}$ have the topology $\mathscr{T} = \{U \subseteq \mathbb{R} : U = \mathbb{R} \text{ or } [0, 1] \cap U = \varnothing\}$. Prove that X is compact.
6. Show that if A and B are compact subsets of a topological space, then $A \cup B$ is compact. Give an example to show that the union of an infinite number of compact sets is not necessarily compact.
7. Let $\mathbb{R}$ have the usual topology. Show that an open interval (a, b) is not compact.
8. Show that if $\mathbb{R}$ has the $\mathscr{C}$ topology, then $\mathbb{R}$ is not compact.

9. Let $\mathbb{R}$ have the $\mathscr{C}$ topology. Show that a subset A of $\mathbb{R}$ is compact iff A contains its greatest lower bound.

10. Prove Theorem 6.1.11.

11. Use Theorem 6.1.20 to determine which of the following sets are compact subsets of $(\mathbb{R}, \mathscr{U})$.

(a) $\mathbb{Z}$ (b) $[0, 1)$ (c) $[0, 1] \cup [5, 8]$
(d) $\{-1, 0, 1\}$ (e) $[-50, 50]$ (f) $\mathbb{Q}$
(g) $\mathbb{R} - \mathbb{Q}$ (h) $\{3n + 1 : n \in \mathbb{Z}^+\}$

12. Prove that $(\mathbb{R}, \mathscr{H})$ is Hausdorff.

13. Which of the following topological spaces are Hausdorff? If a space is not Hausdorff, explain why it is not.

(a) $(\mathbb{R}, \mathscr{I})$
(b) $(\mathbb{R}, \mathscr{D})$
(c) $([0, 1], \mathscr{U}_{[0,1]})$
(d) $(\mathbb{R}, \mathscr{T})$ where $\mathscr{T} = \{U \subseteq \mathbb{R} : U = \mathbb{R} \text{ or } 2 \notin U\}$
(e) $(\mathbb{R}, \mathscr{S})$ where $\mathscr{S} = \{V \subseteq \mathbb{R} : V = \varnothing \text{ or } 2 \in V\}$.

14. Let $(X, \mathscr{T})$ be a topological space with $A \subseteq B \subseteq X$. Prove that if A is $\mathscr{T}$-closed, then A is $\mathscr{T}_B$-closed. (This result was used in the proof of the Heine–Borel Theorem.)

15. Let $(X, \mathscr{T})$ be a topological space with $A \subseteq B \subseteq X$. Prove that if A is $\mathscr{T}_B$-compact, then A is $\mathscr{T}$-compact. (This result was also used in the proof of the Heine–Borel Theorem.)

16. Prove the Bolzano–Weierstrass Theorem (Theorem 6.1.22).

17. For both the topologies $\mathscr{H}$ and $\mathscr{C}$, describe the bounded infinite subsets of $\mathbb{R}$ which do not have limit points.

18. Give an example of each of the following:

(a) a topology $\mathscr{T}$ on $\mathbb{R}$ for which there is a closed and bounded subset A of $\mathbb{R}$ that is not compact.

(b) a topology $\mathscr{S}$ on $\mathbb{R}$ for which there is a compact subset A of $\mathbb{R}$ that is neither closed nor bounded.

19. Prove that if $(X, \mathscr{T})$ is a compact topological space and $\mathscr{S}$ is any topology on X that is coarser than $\mathscr{T}$, then $(X, \mathscr{S})$ is compact.

20. Give an example to show that a set X can have topologies $\mathscr{T}$ and $\mathscr{F}$ with $\mathscr{F}$ finer than $\mathscr{T}$, $(X, \mathscr{T})$ compact, and $(X, \mathscr{F})$ not compact.

6.2 *Properties of Compact Spaces*

Additional properties and applications of compact topological spaces are developed in this section. In particular, we shall see that the continuous image of a compact space is compact and that products of compact spaces are compact. Properties of continuous functions with compact domains are also investigated.

THEOREM 6.2.1 *Let X and Y be topological spaces with $A \subseteq X$ and let $f: X \to Y$ be a continuous function. If A is compact, then $f(A)$ is compact.*

Proof Let $\mathfrak{C} = \{U_\alpha : \alpha \in \Lambda\}$ be an open cover of $f(A)$. Then

$$f(A) \subseteq \bigcup\{U_\alpha : \alpha \in \Lambda\}.$$

Therefore

$$A \subseteq f^{-1}(f(A)) \subseteq f^{-1}(\bigcup\{U_\alpha : \alpha \in \Lambda\}) = \bigcup\{f^{-1}(U_\alpha) : \alpha \in \Lambda\}.$$

Hence the collection $\mathfrak{D} = \{f^{-1}(U_\alpha) : \alpha \in \Lambda\}$ is a cover for A. Since f is continuous, $f^{-1}(U_\alpha)$ is open for each $\alpha \in \Lambda$, and hence $\mathfrak{D}$ is an open cover for A. Because A is compact, $\mathfrak{D}$ has a finite subcover

$$\mathfrak{D}' = \{f^{-1}(U_{\alpha_i}) : i = 1, 2, \ldots, n\}.$$

Then $A \subseteq \bigcup\{f^{-1}(U_{\alpha_i}) : i = 1, 2, \ldots, n\}$. Thus

$$\begin{aligned} f(A) &\subseteq f(\bigcup\{f^{-1}(U_{\alpha_i}) : i = 1, 2, \ldots, n\}) \\ &= \bigcup\{f(f^{-1}(U_{\alpha_i})) : i = 1, 2, \ldots, n\} \\ &\subseteq \bigcup\{U_{\alpha_i} : i = 1, 2, \ldots, n\}. \end{aligned}$$

Therefore $\{U_{\alpha_i} : i = 1, 2, \ldots, n\}$ is a finite subcover of $\mathfrak{C}$, and hence $f(A)$ is compact. ■

The next result is an immediate consequence of Theorem 6.2.1. The proof is left as an exercise.

THEOREM 6.2.2 *Compactness is a topological property.*

THEOREM 6.2.3 *Let X be a compact topological space and let Y be a Hausdorff topological space. If $f: X \to Y$ is a continuous one-to-one onto function, then f is a homeomorphism.*

Proof We must show that f^{-1} is continuous. Let A be a closed subset of X. Since X is compact, it follows that A is compact. Then $(f^{-1})^{-1}(A) = f(A)$ is compact because f is continuous. Since Y is Hausdorff, we have that $f(A)$ is closed. Therefore f^{-1} is continuous. ■

Definition 6.2.4

Let X be a set. A function $f: X \to \mathbb{R}$ is said to be *bounded* if $f(X)$ is a bounded subset of $\mathbb{R}$.

THEOREM 6.2.5 *Let $\mathbb{R}$ have the usual topology and let X be a compact topological space. If $f: X \to \mathbb{R}$ is continuous, then f is bounded.*

Proof Since f is continuous and X is compact, $f(X)$ is a compact subset of $\mathbb{R}$. By the Heine–Borel Theorem, $f(X)$ is bounded. ■

If X is a set, then a function $f: X \to \mathbb{R}$ is said to assume its maximum and minimum if there exist $x_1, x_2 \in X$ such that $f(x_1) \leqq f(x) \leqq f(x_2)$ for all $x \in X$.

Example 6.2.6

Assume $\mathbb{R}$ has the usual topology. Let $f: [-1, 1] \to \mathbb{R}$ be given by $f(x) = x^2$. Then for each $x \in [-1, 1]$, $f(0) \leqq f(x) \leqq f(1)$. Hence f assumes its maximum and minimum. Let $g: (-1, 1) \to \mathbb{R}$ be defined by $f(x) = x^3$. The function g is bounded since $g((-1, 1)) \subseteq [-1, 1]$, but g does not assume either its maximum or its minimum. The reason for the difference in the behavior of these two functions is that $[-1, 1]$ is compact while $(-1, 1)$ is not.

THEOREM 6.2.7 *Let X be a compact nonempty topological space and let $\mathbb{R}$ have the usual topology. If $f: X \to \mathbb{R}$ is a continuous function, then f assumes its maximum and minimum.*

Proof By Theorem 6.2.5, $f(X)$ is a bounded subset of $\mathbb{R}$. Therefore $f(X)$ has a least upper bound m and a greatest lower bound n. Lemmas 5.1.9 and 5.1.10 imply that m and n are in $\mathrm{Cl}(f(X))$. The set $f(X)$ is a compact subset of $\mathbb{R}$ and hence by the Heine–Borel Theorem $f(X)$ is closed. Therefore $\mathrm{Cl}(f(X)) = f(X)$ and $m, n \in f(X)$. Hence there exist $x_1, x_2 \in X$ such that $m = f(x_1)$ and $n = f(x_2)$, and we have that $f(x_2) \leqq f(x) \leqq f(x_1)$ for all $x \in X$. ■

Compact subsets of $(\mathbb{R}, \mathscr{U})$ play a special role with respect to continuity. It turns out that continuous functions with compact domains satisfy the stronger type of continuity given in the following definition.

Definition 6.2.8

Assume $\mathbb{R}$ has the usual topology and let $A \subseteq \mathbb{R}$. A function $f: A \to \mathbb{R}$ is said to be *uniformly continuous* provided that given $\varepsilon > 0$, there exists $\delta > 0$ such that for all $x, y \in A$ $|x - y| < \delta \Rightarrow |f(x) - f(y)| < \varepsilon$.

Note that for a continuous function the value of δ may depend upon both x and ε. However, for a uniformly continuous function the value of δ depends only upon ε. Obviously uniform continuity implies continuity.

In the following theorem the usual ε-δ characterization of continuity from calculus is used. Of course, in the context of the real numbers with the $\mathscr{U}$ topology, this characterization is equivalent to our usual notion of continuity (see Theorems 2.1.8 and 3.2.15).

THEOREM 6.2.9 *Assume $\mathbb{R}$ has the usual topology and let A be a compact subset of $\mathbb{R}$. If $f: A \to \mathbb{R}$ is a continuous function, then f is uniformly continuous.*

Proof Let $\varepsilon > 0$. For each $x \in A$ there exist $\delta_x > 0$ such that for all $y \in A$, $|x - y| < \delta_x$ implies that $|f(x) - f(y)| < \varepsilon/2$. For each $x \in A$ let $U_x = (x - \frac{1}{2}\delta_x, x + \frac{1}{2}\delta_x)$. Obviously the collection $\{U_x : x \in A\}$ is an open cover of A. Therefore A is covered by a finite collection of the intervals, U_{x_1}, $U_{x_2}, \ldots, U_{x_n}$. Let $\delta = \min.\{\frac{1}{2}\delta_{x_1}, \frac{1}{2}\delta_{x_2}, \ldots, \frac{1}{2}\delta_{x_n}\}$. Assume x, $y \in A$ and

$|x - y| < \delta$. Now there exists $i \in \{1, 2, \ldots, n\}$ for which $x \in U_{x_i}$. Hence $|x_i - x| < \frac{1}{2}\delta_{x_i}$. Using the triangle inequality we obtain

$$|x_i - y| = |(x_i - x) + (x - y)| \leqq |x_i - x| + |x - y| < \tfrac{1}{2}\delta_{x_i} + \delta$$
$$\leqq \tfrac{1}{2}\delta_{x_i} + \tfrac{1}{2}\delta_{x_i} = \delta_{x_i}.$$

Thus we have that $|x_i - y| < \delta_{x_i}$. Therefore $|f(x_i) - f(y)| < \varepsilon/2$. Since $|x_i - x| < \frac{1}{2}\delta_{x_i}$, it follows that $|f(x_i) - f(x)| < \varepsilon/2$. Using the triangle inequality again we obtain

$$|f(x) - f(y)| = |(f(x) - f(x_i)) + (f(x_i) - f(y))|$$
$$\leqq |f(x) - f(x_i)| + |f(x_i) - f(y)| < (\varepsilon/2) + (\varepsilon/2) = \varepsilon.$$

This proves that f is uniformly continuous. ■

Some of the continuous functions encountered in elementary calculus have natural domains which are compact with respect to the $\mathscr{U}$ topology and hence are uniformly continuous. For example, both the arcsine and the arccosine functions have $[-1, 1]$ as a domain and thus both are uniformly continuous. The function $f(x) = \sqrt{4 - x^2}$ has $[-2, 2]$ as a domain and hence is uniformly continuous with respect to the $\mathscr{U}$ topology.

Our next goal is to prove that the product of two compact spaces is compact. The following lemma will simplify the proof.

LEMMA 6.2.10 *A topological space is compact* iff *every cover consisting of basic open sets (from a fixed base) has a finite subcover.*

The proof is left as an exercise.

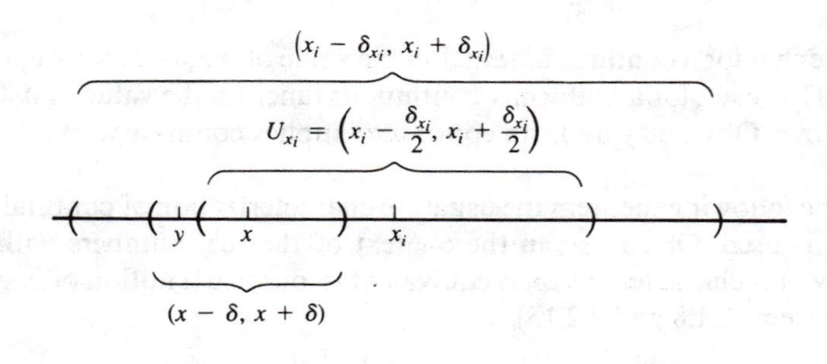

Figure 6.2.1

THEOREM 6.2.11 *Let X and Y be nonempty topological spaces. The product space $X \times Y$ is compact* iff *both X and Y are compact.*

Proof ($\Leftarrow$) Assume X and Y are compact. Let $\mathfrak{C}$ be an open cover of $X \times Y$ consisting of basic open sets of the form $U \times V$ where U is an open subset of X and V is an open subset of Y. By Lemma 6.2.10 it is sufficient to show that $\mathfrak{C}$ has finite subcover.

There are two steps in the proof. First we shall show that for each $x \in X$ there exists an open subset U_x containing x such that $U_x \times Y$ is contained in the union of a finite number of sets in $\mathfrak{C}$. Then we shall show that $X \times Y$ can be expressed as the union of a finite number of sets of the form $U_x \times Y$. The set $U_x \times Y$ can be thought of as a "vertical strip" in the product space $X \times Y$ (see Figure 6.2.2).

Let $x \in X$. Then for each $y \in Y$ there exists a set $U_y \times V_y \in \mathfrak{C}$ containing (x, y). The collection $\{V_y : y \in Y\}$ is an open cover of Y. Since Y is compact, this collection has a finite subcover $\{V_{y_i} : i = 1, 2, \ldots, n\}$. Let $U_x = \bigcap\{U_{y_i} : i = 1, 2, \ldots, n\}$. Then

$$\begin{aligned}\{x\} \times Y \subseteq U_x \times Y = U_x \times (\textstyle\bigcup\{V_{y_i} : i = 1, 2, \ldots, n\}) \\ = \textstyle\bigcup\{U_x \times V_{y_i} : i = 1, 2, \ldots, n\} \\ \subseteq \textstyle\bigcup\{U_{y_i} \times V_{y_i} : i = 1, 2, \ldots, n\}.\end{aligned}$$

Therefore for each $x \in X$ there is a set U_x such that $\{x\} \times Y \subseteq U_x \times Y$ and such that $U_x \times Y$ is contained in the union of a finite number of sets in $\mathfrak{C}$.

Next note that the collection $\{U_x : x \in X\}$ is an open cover of X. Since X is compact, this collection has a finite subcover $\{U_{x_i} : i = 1, 2, \ldots, m\}$. Then $X \times Y = (\bigcup\{U_{x_i} : i = 1, 2, \ldots, m\}) \times Y = \bigcup\{U_{x_i} \times Y : i = 1, 2, \ldots, m\}$. Since for each $i \in \{1, 2, \ldots, m\}$, $U_{x_i} \times Y$ is contained in the union of a finite number of sets in $\mathfrak{C}$, it follows that $X \times Y$ is equal to a union of a finite number of sets in $\mathfrak{C}$. In other words, $\mathfrak{C}$ has a finite subcover, and hence $X \times Y$ is compact.

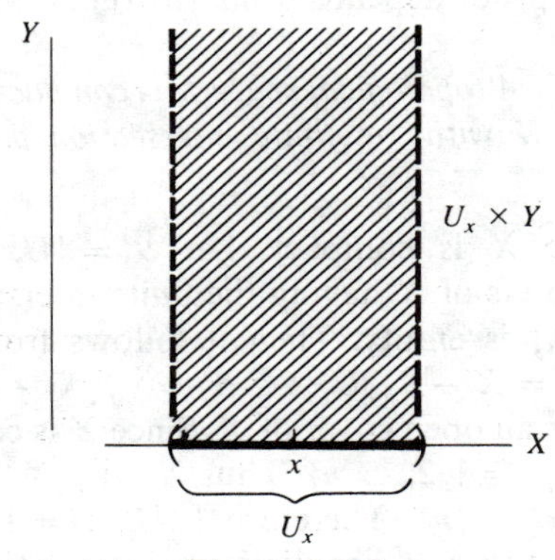

Figure 6.2.2

($\Rightarrow$) Assume that $X \times Y$ is compact. Since the projections $p_1: X \times Y \to X$ and $p_2: X \times Y \to Y$ are continuous, it follows from Theorem 6.2.1 that $X = p_1(X \times Y)$ and $Y = p_2(X \times Y)$ are compact. ■

An induction argument yields the following corollary.

COROLLARY 6.2.12 *Let $X_1, X_2, \ldots, X_n$ be nonempty topological spaces. The product space $X_1 \times X_2 \times \cdots \times X_n$ is compact* iff *X_i is compact for each $i \in \{1, 2, \ldots, n\}$.*

Theorem 6.2.11 can actually be extended to an arbitrary collection of topological spaces. We state this result, which was proved by A. N. Tychonoff in 1935, and omit the proof since it is beyond the scope of this text.

THEOREM 6.2.13 (The Tychonoff Theorem) *Let $\{X_\alpha : \alpha \in \Lambda\}$ be a collection of nonempty topological spaces. The product space $\mathbf{X}\{X_\alpha : \alpha \in \Lambda\}$ is compact* iff *X_α is compact for each $\alpha \in \Lambda$.*

The following definition leads to a useful characterization of compactness.

Definition 6.2.14

A collection $\mathscr{S}$ of subsets of a set X is said to have the *finite intersection property* provided that every finite subcollection of $\mathscr{S}$ has nonempty intersection.

Example 6.2.15

The collection of open intervals $\{(0, 1/n) : n \in \mathbb{Z}^+\}$ has the finite intersection property. Note that the intersection of the entire collection is empty.

The next theorem characterizes compactness in terms of the finite intersection property. The proof depends primarily on DeMorgan's Law.

THEOREM 6.2.16 *A topological space X is compact* iff *every collection of closed subsets of X with the finite intersection property has nonempty intersection.*

Proof ($\Rightarrow$) Assume X is compact. Let $\mathscr{S} = \{U_\alpha : \alpha \in \Lambda\}$ be a collection of closed subsets of X having the finite intersection property. Suppose $\bigcap\{U_\alpha : \alpha \in \Lambda\}$ is empty. Then it follows from DeMorgan's Law that $X = X - \varnothing = X - \bigcap\{U_\alpha : \alpha \in \Lambda\} = \bigcup\{X - U_\alpha : \alpha \in \Lambda\}$. Therefore $\{X - U_\alpha : \alpha \in \Lambda\}$ is an open cover of X. Since X is compact, there is finite subcover $\{X - U_{\alpha_i} : i = 1, 2, \ldots, n\}$. Thus $X = \bigcup\{X - U_{\alpha_i} : i = 1, 2, \ldots, n\} = X - \bigcap\{U_{\alpha_i} : i = 1, 2, \ldots, n\}$. Therefore $\bigcap\{U_{\alpha_i} : i = 1, 2, \ldots, n\} = \varnothing$ which contradicts the fact that $\mathscr{S}$ has the finite intersection property. Thus $\mathscr{S}$ must have nonempty intersection.

The proof of the converse uses DeMorgan's Law in an analogous manner and is left as an exercise. ■

The characterization of compactness given in the preceding theorem yields an easy proof that $(\mathbb{R}, \mathscr{U})$ is not compact.

Example 6.2.17

The collection $\mathscr{S} = \{[a, +\infty) : a \in \mathbb{R}^+\}$ is a collection of closed subsets of $(\mathbb{R}, \mathscr{U})$ with the finite intersection property. However, $\mathscr{S}$ has empty intersection and therefore Theorem 6.2.16 implies that $(\mathbb{R}, \mathscr{U})$ is not compact. (The details of this example are left as an exercise.)

The following result, known as Cantor's Theorem of Deduction, follows easily from Theorem 6.2.16.

Definition 6.2.18

A sequence of sets $\{U_n : n \in \mathbb{Z}^+\}$ is said to be *nested* provided that for each $n \in \mathbb{Z}^+$, $U_{n+1} \subseteq U_n$.

THEOREM 6.2.19 (Cantor's Theorem of Deduction) *If* $\mathscr{S} = \{U_n : n \in \mathbb{Z}^+\}$ *is a nested sequence of closed and bounded, nonempty subsets of* $(\mathbb{R}, \mathscr{U})$, *then* $\mathscr{S}$ *has nonempty intersection.*

The proof is left as an exercise.

Exercises 6.2

1. Let $\mathbb{R}$ have the usual topology and let $f: \mathbb{R} \to \mathbb{R}$ be a continuous function. Prove that if A is a closed and bounded subset of $\mathbb{R}$, then $f(A)$ is also closed and bounded.
2. Let X and Y be topological spaces with $A \subseteq X$ and let $f: X \to Y$ be an open function. Prove that if $\mathfrak{C} = \{U_\alpha : \alpha \in \Lambda\}$ is an open cover of A, then $\{f(U_\alpha) : \alpha \in \Lambda\}$ is an open cover of $f(A)$.
3. Let X and Y be topological spaces and let $f: X \to Y$ be a one-to-one open function from X onto Y. Prove that if B is a compact subset of Y, then $f^{-1}(B)$ is a compact subset of X.
4. Let X and Y be topological spaces and let $f: X \to Y$ be a one-to-one onto function with a continuous inverse. Prove that if X is Hausdorff and Y is compact, then f is a homeomorphism.
5. Let X be a compact topological space and let Y be a Hausdorff topological space. Show that if $f: X \to Y$ is a continuous function, then the inverse image of each compact subset of Y is a compact subset of X.
6. Let X and Y be topological spaces with $x \in X$. Use Theorem 6.2.1 to prove that if Y is compact, then $\{x\} \times Y$ is a compact subset of the product space $X \times Y$.
7. Assume each of the sets in the following products has the $\mathscr{U}$-relative topology. Determine which of the products are compact.

   (a) $[0, 1] \times [0, 100]$ **(b)** $(0, 1) \times [1, 2]$

(c) $\{0\} \times ([1,2] \cup [5,8])$ (d) $\{0,1,2\} \times \{3,4,5\}$
(e) $[0,+\infty) \times [1,2]$

8. Prove Theorem 6.2.2.

9. Prove Lemma 6.2.10.

10. Prove the collection $\mathscr{S}$ in Example 6.2.17 has the finite intersection property and that the intersection of the entire collection is empty.

11. Complete the proof of Theorem 6.2.16.

12. Prove Theorem 6.2.19.

Review Exercises 6

Mark each of the following statements true or false. Briefly explain each true statement and find a counterexample for each false statement.

1. Every finite topological space is compact.

2. Every compact topological space is finite.

3. An interval is a compact subset of $(\mathbb{R}, \mathscr{U})$.

4. A closed interval is a compact subset of $(\mathbb{R}, \mathscr{U})$.

5. The compact subsets of $(\mathbb{R}, \mathscr{U})$ are precisely the closed intervals and singleton sets.

6. The compact subsets of $(\mathbb{R}, \mathscr{U})$ are precisely the closed sets which are contained in a closed interval.

7. Every subset of a compact topological space is compact.

8. Every Hausdorff subset of a compact topological space is compact.

9. Every closed subset of a compact topological space is compact.

10. Every compact subset of a Hausdorff topological space is closed.

11. Every compact subset of a topological space is closed.

12. The union of two compact subsets of a topological space is compact.

13. The union of any collection of compact subsets of a topological space is compact.

14. If $\mathbb{R}$ has the usual topology, then any continuous function from $[0,1]$ to $\mathbb{R}$ assumes its maximum and minimum.

15. If $\mathbb{R}$ has the usual topology, then any continuous function from $[0,1)$ to $\mathbb{R}$ assumes its maximum and minimum.

16. Every compact subet of $(\mathbb{R}, \mathscr{U})$ is connected.

17. Every connected subset of $(\mathbb{R}, \mathscr{U})$ is compact.

18. If $\mathbb{R}$ has the usual topology, then any continuous function from $(0,1]$ to $\mathbb{R}$ is uniformly continuous.

19. If $\mathbb{R}$ has the usual topology, then any continuous function from $[0,1]$ to $\mathbb{R}$ is uniformly continuous.

20. Let $\mathbb{R}$ have the usual topology and let $A \subseteq \mathbb{R}$. If A is not compact, then there is no uniformly continuous function from A to $\mathbb{R}$.

7

Separation Properties

7.1 *T_0-, T_1-, and T_2-Spaces*

When Hausdorff first introduced the concept of a topological space, he considered only T_2-spaces. However, several similar properties have since been developed and have turned out to be quite useful. In this section some of these properties are investigated and additional properties of Hausdorff spaces are developed. For completeness the definition of a Hausdorff space is restated.

Definition 7.1.1

Let X be a topological space.

(a) The space X is said to be a *T_0-space* if for any two points x and y in X there exists an open set containing x but not y or there exists an open set containing y but not x.

(b) The space X is said to be a *T_1-space* if for any two points x and y in X there exists an open set containing x but not y and there exists an open set containing y but not x.

(c) The space X is said to be a *T_2-space* or a *Hausdorff space* if for any two points x and y in X there exist two disjoint open sets containing x and y, respectively.

Note the use of the word "or" in the definition of a T_0-space and the word "and" in the definition of a T_1-space. If X is a T_0-space, there is an open set containing one of the points but not the other. If X is a T_1-space, there are two open sets, one containing x but not y and the other containing y but not x.

However, the T_1-property does not require the two sets to be disjoint. The T_2-property or the Hausdorff property does require the two sets to be disjoint.

All of the above properties are called separation properties because they are concerned with how points in a topological space can be separated by open sets. In other sections of this chapter separation properties involving points and subsets of topological spaces will be investigated. The assumption of a separation property usually eliminates extremely coarse topologies such as the indiscrete topology. Indeed, some separation condition is usually necessary in order for a space to have useful properties. For example, we shall see in Chapter 8 that the Hausdorff property ensures that sequences do not converge to more than one point.

The T_i-terminology for the separation properties, also known as *Trennungsaxiomen*, was developed by Heinrich Tietze in 1923. The T_0- and T_1-properties were introduced by A. N. Kolmogorov, and Frechet (in 1907), respectively. It was in 1914 that Hausdorff incorporated the T_2-property into his definition of a topological space.

The proofs of the implications given in the following theorem are straightforward and are left as exercises.

THEOREM 7.1.2 *Let X be a topological space.*

(a) *If X is a T_1-space, then X is a T_0-space.*

(b) *If X is a T_2-space, then X is a T_1-space.*

The following examples show that the implications in Theorem 7.1.2 cannot be reversed.

Example 7.1.3

Let $X = \mathbb{R}$ have the $\mathscr{C}$ topology and let $x, y \in X$. Assume that $x < y$. If $U = (x, +\infty)$, then U is an open set containing y but not x. However, there is no open set containing x but not y. Thus X is a T_0-space but not a T_1-space.

Example 7.1.4

Let $X = \mathbb{R}$ have the finite complement topology. Let x and y be two points in X. If $U = X - \{y\}$, then U is an open set containing x but not y. Therefore X is a T_1-space. To see that X is not a T_2-space, suppose that A and B are disjoint open subsets of X. Then $X = X - (A \cap B) = (X - A) \cup (X - B)$. This is a contradiction since X is infinite and both $X - A$ and $X - B$ are finite. Since there are no disjoint open sets, X cannot be a T_2-space.

Recall that both $(\mathbb{R}, \mathscr{U})$ and $(\mathbb{R}, \mathscr{H})$ are T_2-spaces and hence T_i-spaces for $i = 0$ or 1.

THEOREM 7.1.5 *For each $i \in \{0, 1, 2\}$, the T_i-property is a topological property.*

Proof We shall prove the theorem for T_0-spaces and leave the other two proofs as exercises.

Let X be a T_0-space and let Y be a space that is homeomorphic to X. Let $f: X \to Y$ be a homeomorphism. Assume y_1 and y_2 are distinct points in Y. Since f is one-to-one and onto, there exist distinct points x_1 and x_2 in X such that $f(x_1) = y_1$ and $f(x_2) = y_2$. Because X is a T_0-space, there exists an open subset U of X containing one of the points x_1 or x_2 but not the other. Then obviously $f(U)$ contains one of the points y_1 or y_2. Since f is one-to-one, $f(U)$ does not contain the other point. Finally, since f is an open function, $f(U)$ is an open set. Therefore Y is a T_0-space. ■

The following two theorems give useful characterizations of the T_0- and T_1-properties in terms of the closures of singleton sets.

THEOREM 7.1.6 *A topological space X is a T_0-space* iff *for any two points x and y in X* $\mathrm{Cl}(\{x\}) \neq \mathrm{Cl}(\{y\})$.

Proof ($\Rightarrow$) Assume X is a T_0-space. Let x and y be two points in X. Then there exists an open set U containing one of the points, x or y but not the other. Assume U contains x but not y. Since $x \in U$ and $U \cap \{y\} = \varnothing$, it follows that $x \notin \mathrm{Cl}(\{y\})$ and hence $\mathrm{Cl}(\{x\}) \neq \mathrm{Cl}(\{y\})$. Similarly if U contains y but not x, an analogous argument shows that $y \notin \mathrm{Cl}(\{x\})$ and hence $\mathrm{Cl}(\{x\}) \neq \mathrm{Cl}(\{y\})$. Thus in either case $\mathrm{Cl}(\{x\}) \neq \mathrm{Cl}(\{y\})$.

($\Leftarrow$) Assume that for any two points x and y in X, $\mathrm{Cl}(\{x\}) \neq \mathrm{Cl}(\{y\})$. Let x and y be two points in X. Note that if $x \in \mathrm{Cl}(\{y\})$ and $y \in \mathrm{Cl}(\{x\})$, then $\mathrm{Cl}(\{x\}) \subseteq \mathrm{Cl}(\{y\})$ and $\mathrm{Cl}(\{y\}) \subseteq \mathrm{Cl}(\{x\})$. That is, if $x \in \mathrm{Cl}(\{y\})$ and $y \in \mathrm{Cl}(\{x\})$ then $\mathrm{Cl}(\{x\}) = \mathrm{Cl}(\{y\})$. Thus we must have that either $x \notin \mathrm{Cl}(\{y\})$ or $y \notin \mathrm{Cl}(\{x\})$. Assume that $x \notin \mathrm{Cl}(\{y\})$. Then there exists an open set U containing x such that $U \cap \{y\} = \varnothing$. That is, U contains x but not y. Similarly if $y \notin \mathrm{Cl}(\{x\})$, there is an open set V containing y but not x. This proves that X is a T_0-space. ■

THEOREM 7.1.7 *A topological space X is a T_1-space* iff *for each $x \in X$, $\{x\}$ is a closed set.*

Proof ($\Rightarrow$) Let X be a T_1-space and let $x \in X$. We shall show that $X - \{x\}$ is open. Let $y \in X - \{x\}$. Because X is a T_1-space, there exists an open set U containing y but not x. Thus $y \in U \subseteq X - \{x\}$ and therefore $X - \{x\}$ is open and $\{x\}$ is closed.

($\Leftarrow$) Assume that for each $x \in X$ the set $\{x\}$ is closed. Let x and y be distinct points in X. Since $\mathrm{Cl}(\{y\}) = \{y\}$, $x \notin \mathrm{Cl}(\{y\})$. Therefore there exists an open set U containing x such that $U \cap \{y\} = \varnothing$. That is, U is an open set containing x but not y. An analogous argument shows that $y \notin \mathrm{Cl}(\{x\})$ and that there exists an open set V containing y but not x. This proves X is a T_1-space. ■

The proof of the following corollary is left as an exercise.

COROLLARY 7.1.8 *A topological space is a* T_1*-space* iff *every finite subset is closed.*

Many of the topological properties we have studied are not inherited by subspaces. For example, a subspace of a compact space is not necessarily compact. However, as we see in the next theorem, each of the T_i-properties is inherited by a subspace.

THEOREM 7.1.9 *For each* $i \in \{0, 1, 2\}$*, if* X *is a* T_i*-space and* $A \subseteq X$*, then* A *is a* T_i*-space.*

The proof is left as an exercise.

Products of T_i-spaces are also T_i-spaces for each $i \in \{0, 1, 2\}$.

THEOREM 7.1.10 *Let* X *and* Y *be nonempty topological spaces. For each* $i \in \{0, 1, 2\}$*, the product space* $X \times Y$ *is a* T_i*-space* iff *both* X *and* Y *are* T_i*-spaces.*

Proof We shall prove the theorem for T_2-spaces and leave the other proofs as exercises.

($\Leftarrow$) Assume both X and Y are T_2-spaces. Let (x_1, y_1) and (x_2, y_2) be distinct points in $X \times Y$. The two points must differ in at least one coordinate. Without loss of generality, assume $x_1 \neq x_2$. Since X is a T_2-space, there exist disjoint open subsets U and V of X for which $x_1 \in U$ and $x_2 \in V$. Then $U \times Y$ and $V \times Y$ are disjoint open subsets $X \times Y$ containing (x_1, y_1) and (x_2, y_2), respectively (see Figure 7.1.1). Therefore $X \times Y$ is a T_2-space.

($\Rightarrow$) Assume $X \times Y$ is a T_2-space. We shall show that X is a T_2-space. (The proof that Y is a T_2-space is completely analogous.)

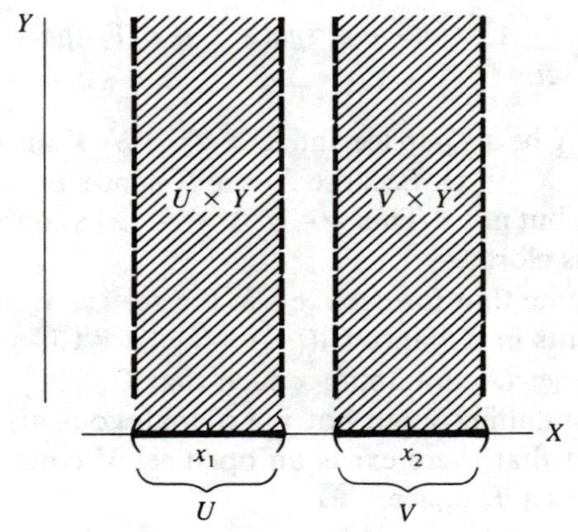

Figure 7.1.1

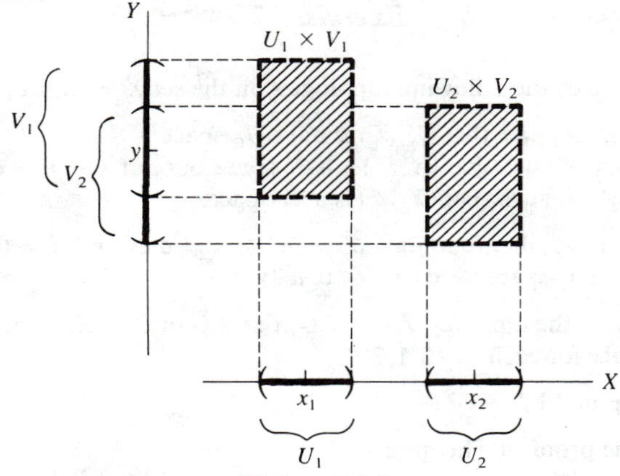

Figure 7.1.2

Suppose x_1 and x_2 are distinct points in X. Let $y \in Y$. Then (x_1, y) and (x_2, y) are distinct points in $X \times Y$. Since $X \times Y$ is a T_2-space, there exist disjoint basic open sets $U_1 \times V_1$ and $U_2 \times V_2$ containing (x_1, y) and (x_2, y), respectively (see Figure 7.1.2). Clearly $x_1 \in U_1$, $x_2 \in U_2$. To see that U_1 and U_2 are disjoint, suppose $z \in U_1 \cap U_2$. Then $(z, y) \in (U_1 \times V_1) \cap (U_2 \times V_2)$ which is a contradiction because the two sets are disjoint. Thus X is a T_2-space. ■

For any $i \in \{0, 1, 2\}$, continuous images of T_i-spaces are not necessarily T_i-spaces. For example, let $(X, \mathscr{T})$ be any T_2-space containing two or more points and let $\mathscr{I}$ be the indiscrete topology on X. The identity function from $(X, \mathscr{T})$ onto $(X, \mathscr{I})$ is obviously continuous. However, $(X, \mathscr{I})$ is not even a T_0-space.

THEOREM 7.1.11 *Let X and Y be topological spaces and let $f\colon X \to Y$ and $g\colon X \to Y$ be continuous functions. If Y is a T_2-space, then the set $A = \{x \in X : f(x) = g(x)\}$ is a closed subset of X.*

Proof We shall show that $X - A$ is open. Suppose $x \in X - A$. Then $f(x) \neq g(x)$. Since Y is a T_2-space, there exist disjoint open subsets U and V of Y containing $f(x)$ and $g(x)$, respectively. The continuity of f and g implies that $f^{-1}(U)$ and $g^{-1}(V)$ are open subsets of X. Then the set $f^{-1}(U) \cap g^{-1}(V)$ is an open subset of X such that $x \in f^{-1}(U) \cap g^{-1}(V) \subseteq X - A$. (To see this, note that if $z \in f^{-1}(U) \cap g^{-1}(V)$, then $f(z) \in U$ and $g(z) \in V$. Since U and V are disjoint, $f(z) \neq g(z)$ and therefore $z \notin A$.) Therefore $X - A$ is open and hence A is closed. ■

COROLLARY 7.1.12 *Let X and Y be topological spaces and let $f\colon X \to Y$ and $g\colon X \to Y$ be continuous functions. If Y is a T_2-space and $f(x) = g(x)$ for all x in a dense subset of X, then $f = g$.*

The proof is left as an exercise.

Exercises 7.1

1. Give examples of the following topologies on the set $X = \{a, b, c\}$:

   (a) a topology $\mathscr{T}$ such that $(X, \mathscr{T})$ is not a T_0-space
   (b) a topology $\mathscr{S}$ such that $(X, \mathscr{S})$ is a T_0-space but not a T_1-space
   (c) a topology $\mathscr{F}$ such that $(X, \mathscr{F})$ is a T_1-space.

2. Let $X = \mathbb{R}$ have the topology $\mathscr{T} = \{U \subseteq X : 1 \in U \text{ or } U = \varnothing\}$. Determine whether X is a T_i-space for each $i \in \{0, 1, 2\}$.

3. Let $X = \mathbb{R}$ have the topology $\mathscr{S} = \{U \subseteq X : 1 \notin U \text{ or } U = X\}$. Determine whether X is a T_i-space for each $i \in \{0, 1, 2\}$.

4. Prove Theorem 7.1.2.

5. Complete the proof of Theorem 7.1.5.

6. Let X and Y be topological spaces. Prove that if Y is a T_1-space and $f: X \to Y$ is a continuous one-to-one function, then X is a T_1-space.

7. Prove Corollary 7.1.8.

8. Prove Theorem 7.1.9.

9. Let X and Y be topological spaces. Prove that if Y is a T_2-space and $f: X \to Y$ is a continuous one-to-one function, then X is a T_2-space.

10. Prove Theorem 7.1.10 for T_0-spaces and T_1-spaces.

11. Prove Corollary 7.1.12.

12. Let X be a T_1-space with $A \subseteq X$. Prove that if x is a limit point of A and U is an open set containing x, then $U \cap A$ is an infinite set.

13. Let X be a T_1-space. Prove that if A is a finite subset of X, then A does not have a limit point.

14. Let X and Y be topological spaces and let $f: X \to Y$ be a function. The *graph* of f is the set $G(f) = \{(x, y) \in X \times Y : y = f(x)\}$. Prove that if f is continuous and Y is a T_2-space, then $G(f)$ is a closed subset of the product space $X \times Y$.

15. Prove that if $(X, \mathscr{T})$ is a finite T_1-space, then $\mathscr{T}$ is the discrete topology.

16. Prove that if X is a T_1-space and $A \subseteq X$, then the set of limit points of A, A' is a closed subset of X.

7.2 Regular Spaces

A separation property involving points and subsets of a topological space is developed in this section. This property, called regularity, was studied first by Vietoris in 1921. We shall see that for T_1-spaces this separation property is more restrictive than the Hausdorff condition.

Definition 7.2.1

A topological space X is said to be *regular* if, for every closed subset F of X and every point $x \in X - F$, there exist disjoint open sets U and V such that $F \subseteq U$ and $x \in V$.

Example 7.2.2

If $X = \{a, b, c\}$ has the topology $\mathscr{T} = \{X, \varnothing, \{a, b\}, \{c\}\}$, then X is a regular space. Note that the only proper nonempty closed subsets of X are $\{c\}$ and $\{a, b\}$. For $F = \{c\}$ and $x = a$ or b, let $U = \{c\}$ and $V = \{a, b\}$. Similarly, for $F = \{a, b\}$ and $x = c$, let $U = \{a, b\}$ and $V = \{c\}$.

Example 7.2.3

Let X be an infinite set and let $a \in X$. Let $\mathscr{T} = \{U \subseteq X : X - U$ is finite or $a \in X - U\}$. The space $(X, \mathscr{T})$, called the Fort space, is regular. To see this, let F be a closed subset of X and let $x \in X - F$. Suppose $x \neq a$. Then $\{x\}$ is open. The set $X - \{x\}$ is also open because it has a finite complement. Thus if $U = X - \{x\}$ and $V = \{x\}$, then U and V are disjoint open sets containing F and x, respectively. Suppose $x = a$. Then $a \in X - F$ and hence F is open. Therefore if $U = F$ and $V = X - F$, then U and V are disjoint open sets containing F and x, respectively.

Example 7.2.4

If $X = \mathbb{R}$ has the $\mathscr{C}$ topology, then X is not regular. Note that $F = (-\infty, 0]$ is a closed set and that $1 \notin F$, but any open set containing F must also contain 1.

Example 7.2.5

Let $X = \mathbb{R}$ have the topology $\mathscr{T} = \{U \subseteq X : 1 \in U$ or $U = \varnothing\}$. This space is not regular. For example, the closed set $(3, 5)$ and the point 1 cannot be separated by open sets.

Example 7.2.6

Any discrete topological space is obviously regular. Any indiscrete space is also regular since there are no proper nonempty closed sets.

THEOREM 7.2.7 *The space $(\mathbb{R}, \mathscr{U})$ is regular.*

Proof Let F be a closed subset of $\mathbb{R}$ and let $x \in \mathbb{R} - F$. Since $F = \mathrm{Cl}(F)$, $x \notin \mathrm{Cl}(F)$. Hence there exists an open set W such that $x \in W$ and $W \cap F = \varnothing$. Let $\varepsilon > 0$ such that $(x - \varepsilon, x + \varepsilon) \subseteq W$. Let $U = \left(x - \dfrac{\varepsilon}{2}, x + \dfrac{\varepsilon}{2}\right)$ and let $V = \left(-\infty, x - \dfrac{\varepsilon}{2}\right) \cup \left(x + \dfrac{\varepsilon}{2}, +\infty\right)$. It is clear that U and V are disjoint open sets and that $x \in U$. Since $F \subseteq \mathbb{R} - W \subseteq \mathbb{R} - (x - \varepsilon, x + \varepsilon) \subseteq V$, we have that $F \subseteq V$. This proves that $(\mathbb{R}, \mathscr{U})$ is regular. ■

THEOREM 7.2.8 *The space $(\mathbb{R}, \mathscr{H})$ is regular.*

The proof is left as an exercise.

The following theorems give some of the properties of regular spaces. The first theorem relates regularity to the closures of singleton sets.

THEOREM 7.2.9 *If X is a regular topological space and x and y are two points in X, then either* $\mathrm{Cl}(\{x\}) = \mathrm{Cl}(\{y\})$ *or* $\mathrm{Cl}(\{x\}) \cap \mathrm{Cl}(\{y\}) = \varnothing$.

The proof is left as an exercise.

The next theorem gives a very useful characterization of regularity. The theorem states that regularity is equivalent to the property of each open set containing a closed neighborhood of each of its points. This property is sometimes easier to use in proofs than the definition of regularity.

THEOREM 7.2.10 *A topological space X is regular* iff, *for each $x \in X$ and each open set U with $x \in U$, there exists an open set V for which $x \in V \subseteq \mathrm{Cl}(V) \subseteq U$.*

Proof ($\Rightarrow$) Assume that X is a regular space. Let $x \in X$ and let U be an open subset of X with $x \in U$. Then $X - U$ is closed and $x \notin X - U$. The regularity of X implies that there exist disjoint open sets V and W such that $x \in V$ and $X - U \subseteq W$. Note that $V \subseteq X - W$ and that $X - W \subseteq U$. Then $x \in V \subseteq \mathrm{Cl}(V) \subseteq \mathrm{Cl}(X - W) = X - W \subseteq U$. Therefore $x \in V \subseteq \mathrm{Cl}(V) \subseteq U$ as desired.

($\Leftarrow$) Assume that, for each $x \in X$ and for each open set U containing x, there exists an open set V such that $x \in V \subseteq \mathrm{Cl}(V) \subseteq U$. Let F be a closed subset of X and let $x \in X - F$. Then there exists an open set V such that $x \in V \subseteq \mathrm{Cl}(V) \subseteq X - F$. Therefore $x \in V$ and $F \subseteq X - \mathrm{Cl}(V)$. The sets V and $X - \mathrm{Cl}(V)$ are obviously open and, since $V \subseteq \mathrm{Cl}(V)$, the sets V and $X - \mathrm{Cl}(V)$ are disjoint. This proves that X is a regular space. ■

Just as for the T_0-, T_1-, and T_2-separation properties, regularity is inherited by subspaces and preserved under products.

THEOREM 7.2.11 *If X is a regular topological space and $A \subseteq X$, then A is regular.*

The proof is straightforward and is left as an exercise.

THEOREM 7.2.12 *Let X and Y be nonempty topological spaces. The product space $X \times Y$ is regular* iff *both X and Y are regular.*

Proof ($\Rightarrow$) Assume $X \times Y$ is regular. We shall show that X is regular. (The proof that Y is regular is similar.) Let $x_0 \in X$ and let U be an open subset of X containing x_0. Let $y_0 \in Y$. Then $U \times Y$ is an open subset of $X \times Y$ containing (x_0, y_0). Since $X \times Y$ is regular, there exists an open subset W of $X \times Y$ such that $(x_0, y_0) \in W \subseteq \mathrm{Cl}(W) \subseteq U \times Y$. Let $V_1 \times V_2$ be a basic open subset of $X \times Y$ such that $(x_0, y_0) \in V_1 \times V_2 \subseteq W$. Then $(x_0, y_0) \in V_1 \times V_2 \subseteq \mathrm{Cl}(V_1 \times V_2) \subseteq \mathrm{Cl}(W) \subseteq U \times Y$. By Theorem 4.1.9 $\mathrm{Cl}(V_1 \times V_2) = \mathrm{Cl}(V_1) \times \mathrm{Cl}(V_2)$. Therefore we have that

$$(x_0, y_0) \in V_1 \times V_2 \subseteq \mathrm{Cl}(V_1) \times \mathrm{Cl}(V_2) \subseteq U \times Y.$$

It follows that $x_0 \in V_1 \subseteq \mathrm{Cl}(V_1) \subseteq U$, which proves that X is regular.

($\Leftarrow$) Assume both X and Y are regular. Let $(x_0, y_0) \in X \times Y$ and let W be an open subset of $X \times Y$ containing (x_0, y_0). There is a basic open set $U \times V$ for which $(x_0, y_0) \in U \times V \subseteq W$. Then $x_0 \in U$ and $y_0 \in V$. Since both X and Y are regular, there exist open subsets U_1 and V_1 of X and Y, respectively, such that $x_0 \in U_1 \subseteq \mathrm{Cl}(U_1) \subseteq U$ and $y_0 \in V_1 \subseteq \mathrm{Cl}(V_1) \subseteq V$. Then $(x_0, y_0) \in U_1 \times V_1 \subseteq \mathrm{Cl}(U_1) \times \mathrm{Cl}(V_1) \subseteq U \times V$. Since Theorem 4.1.9 implies that $\mathrm{Cl}(U_1) \times \mathrm{Cl}(V_1) = \mathrm{Cl}(U_1 \times V_1)$, we have that

$$(x_0, y_0) \in U_1 \times V_1 \subseteq \mathrm{Cl}(U_1 \times V_1) \subseteq U \times V \subseteq W.$$

Therefore $X \times Y$ is regular. ■

It follows easily from the above theorem that the product of any finite number of spaces is regular iff each space is regular.

Since $(\mathbb{R}, \mathscr{U})$ is regular, we have the following corollary.

COROLLARY 7.2.13 *If $\mathbb{R}$ has the usual topology, then for any positive integer n, $\mathbb{R}^n$ is regular.*

Regularity does not imply any of the T_i-properties for $i = 0$, 1, or 2. For example, an indiscrete space with two or more points is obviously not a T_0-space but is trivially a regular space because there are no proper nonempty closed sets. Theorem 7.2.12 is used to construct the following more interesting example of a regular space that is not a T_0-space.

Example 7.2.14

Let $\mathbb{R}$ have the usual topology and let $X = \{0, 1\}$ have the indiscrete topology. Since both $\mathbb{R}$ and X are regular, Theorem 7.2.12 implies that $\mathbb{R} \times X$ is regular. However, since X is not a T_0-space, Theorem 7.1.10 implies that $\mathbb{R} \times X$ is not a T_0-space.

If the T_1-property is combined with regularity, then the resulting condition is stronger than the T_2-property.

Definition 7.2.15

A regular T_1-space is called a *T_3-space.*

Some authors interchange the definitions of the T_3-property and regularity.

The proofs of the following theorems are left as exercises.

THEOREM 7.2.16 *Every T_3-space is a T_2-space.*

THEOREM 7.2.17 *Regularity is a topological property.*

THEOREM 7.2.18 *The T_3-condition is a topological property.*

The last theorem of this section follows easily from Lemma 6.1.18.

THEOREM 7.2.19 *Every compact Hausdorff space is regular.*

Exercises 7.2

1. Find topologies on the set $X = \{a, b, c, d\}$ (different from the discrete and the indiscrete topologies) with the following properties:
   (**a**) a topology $\mathscr{T}$ such that $(X, \mathscr{T})$ is not regular.
   (**b**) a topology $\mathscr{F}$ such that $(X, \mathscr{F})$ is regular.
2. Let $X = \{a, b, c, d, e\}$ and let $\mathscr{T} = \{X, \varnothing, \{a\}, \{b\}, \{a, b\}, \{b, c, d, e\}\}$. Determine whether $(X, \mathscr{T})$ is regular.
3. Let $X = \mathbb{R}$ have the topology $\mathscr{T} = \{U \subseteq X : 1 \notin U \text{ or } U = X\}$. Determine whether $(X, \mathscr{T})$ is regular.
4. Prove that the space $(\mathbb{R}, \mathscr{H})$ is regular (Theorem 7.2.8).
5. Prove Theorem 7.2.9.
6. Prove Theorem 7.2.11.
7. Determine which of the following product spaces are regular:
   (**a**) $X \times Y \times Z$, where $X = \mathbb{R}$ has the usual topology, $Y = \{0, 1\}$ has the indiscrete topology, and $Z = \{0, 1, 2\}$ has the discrete topology
   (**b**) $X \times Y$ where $X = \mathbb{R}$ has the $\mathscr{U}$ topology and $Y = \mathbb{R}$ has the $\mathscr{C}$ topology
   (**c**) $X_1 \times X_2 \times X_3$ where, for each $i \in \{1, 2, 3\}$, $X_i = \mathbb{R}$ with the topology $\mathscr{T}_i = \{U \subseteq \mathbb{R} : i \notin U \text{ or } U = \mathbb{R}\}$.
8. Prove Corollary 7.2.13.
9. Find an example to show that the continuous image of regular space is not necessarily regular.
10. Prove Theorem 7.2.16.
11. Prove Theorem 7.2.17.
12. Prove Theorem 7.2.18.
13. Prove Theorem 7.2.19.

7.3 *Normal Spaces*

In this section we continue the development of increasingly restrictive separation conditions. A separation property called normality, introduced by Vietoris in 1921, is investigated. This concept involves the separation of two subsets of a topological space. We shall see that for T_1-spaces normality is even more restrictive than regularity. Several additional separation properties are discussed briefly.

Definition 7.3.1

A topological space X is said to be *normal* provided that for any two disjoint closed subsets F_1 and F_2 of X, there exist disjoint open sets U and V such that $F_1 \subseteq U$ and $F_2 \subseteq V$.

Normality is a natural extension of regularity. However, normal spaces are not as well behaved as regular spaces. For example, we shall see later that subspaces of normal spaces are not necessarily normal, and products of normal spaces may fail to be normal.

Example 7.3.2

If $X = \mathbb{R}$ has $\mathscr{C}$ topology, then X is normal. Note that there are no disjoint nonempty closed sets. Recall that this space is not regular.

Example 7.3.3

Let $X = \{a, b, c\}$ have the topology $\mathscr{T} = \{X, \varnothing, \{a\}, \{b, c\}\}$. The space $(X, \mathscr{T})$ is normal since the only proper nonempty closed sets are $\{b, c\}$ and $\{a\}$, and these sets are also open.

Example 7.3.4

The Fort space (see Example 7.2.3) is normal. Recall that X is an infinite set with $a \in X$ and $\mathscr{T} = \{U \subseteq X : X - U \text{ is finite or } a \in X - U\}$. To see that $(X, \mathscr{T})$ is normal, let F_1 and F_2 be two disjoint closed sets. Suppose $a \notin F_1$ and $a \notin F_2$. Then F_1 and F_2 are both open. Suppose $a \in F_1$. Since F_1 and F_2 are disjoint, $a \notin F_2$. Then F_2 is open. Since F_2 is closed, $X - F_2$ is also open. Obviously $F_1 \subseteq X - F_2$, $F_2 \subseteq F_2$ and $(X - F_2) \cap F_2 = \varnothing$. The case where $a \in F_1$ and $a \notin F_2$ is analogous.

Example 7.3.5

Let $X = \mathbb{R}$ have the topology $\mathscr{T} = \{U \subseteq \mathbb{R} : 1 \in U \text{ or } U = \varnothing\}$. The space $(X, \mathscr{T})$ is not normal. Note that there do exist disjoint closed sets, for example, the sets $\{2\}$ and $\{3\}$. However, there are no disjoint nonempty open sets.

Example 7.3.6

Every indiscrete space is normal because there are no disjoint nonempty closed sets. Obviously every discrete space is normal since closed sets are also open.

THEOREM 7.3.7 *The space* $(\mathbb{R}, \mathscr{U})$ *is normal.*

Proof Let F_1 and F_2 be two disjoint closed subsets of $\mathbb{R}$. For each $x \in F_1$, since $x \notin F_2$ and F_2 is closed, there exists $\varepsilon_x > 0$ such that $(x - \varepsilon_x, x + \varepsilon_x) \cap F_2 = \varnothing$. Similarly for each $y \in F_2$, there exists $\varepsilon_y > 0$ for which $(y - \varepsilon_y, y + \varepsilon_y) \cap F_1 = \varnothing$. Let $U = \bigcup\{(x - \varepsilon_x/2, x + \varepsilon_x/2) : x \in F_1\}$ and let $V = \bigcup\{(y - \varepsilon_y/2, y + \varepsilon_y/2) : y \in F_2\}$. Clearly U and V are open sets containing F_1 and F_2 respectively. The proof will be completed by showing that U and V are disjoint. Suppose $z \in U \cap V$. Since $z \in U$, there exists $x \in F_1$ such that $z \in (x - \varepsilon_x/2, x + \varepsilon_x/2)$ and hence $|x - z| < \varepsilon_x/2$. Similarly because $z \in V$, there exists $y \in F_2$ such that

$$z \in (y - \varepsilon_y/2, y + \varepsilon_y/2)$$

and thus $|z - y| < \varepsilon_y/2$. By the triangle inequality we have that

$$|x - y| = |(x - z) + (z - y)| \leqq |x - z| + |z - y| < \varepsilon_x/2 + \varepsilon_y/2.$$

Now either $\varepsilon_x \leqq \varepsilon_y$ or $\varepsilon_y \leqq \varepsilon_x$. Suppose $\varepsilon_x \leqq \varepsilon_y$. Then $|x - y| < \varepsilon_y$ and therefore $x \in (y - \varepsilon_y, y + \varepsilon_y)$ which is disjoint from F_1. This is a contradiction since $x \in F_1$. Similarly, if $\varepsilon_y \leqq \varepsilon_x$, we obtain $|x - y| < \varepsilon_x$. Thus $y \in (x - \varepsilon_x, x + \varepsilon_x)$ which is a contradiction since $(x - \varepsilon_x, x + \varepsilon_x)$ is disjoint from F_2 and $y \in F_2$. Therefore $U \cap V = \varnothing$ which proves that $(\mathbb{R}, \mathscr{U})$ is normal. ■

THEOREM 7.3.8 *The space $(\mathbb{R}, \mathscr{H})$ is normal.*

The proof is left as an exercise.

The following two theorems give useful characterizations of normality.

THEOREM 7.3.9 *A topological space X is normal* iff, *for each closed subset F of X and each open set U with $F \subseteq U$, there exists an open set V for which $F \subseteq V \subseteq \mathrm{Cl}(V) \subseteq U$.*

Proof ($\Rightarrow$) Assume that X is normal. Let F be a closed subset of X and let U be an open set containing F. Then F and $X - U$ are disjoint closed sets. The normality of X implies that there exist disjoint open sets V and W such that $F \subseteq V$ and $X - U \subseteq W$. Note that $V \subseteq X - W$ and that $X - W \subseteq U$. Therefore $F \subseteq V \subseteq \mathrm{Cl}(V) \subseteq \mathrm{Cl}(X - W) = X - W \subseteq U$. Thus $F \subseteq V \subseteq \mathrm{Cl}(V) \subseteq U$.

($\Leftarrow$) Assume that, for each closed set F and each open set U containing F, there exists an open set V such that $F \subseteq V \subseteq \mathrm{Cl}(V) \subseteq U$. Let F_1 and F_2 be two disjoint closed sets. Then $X - F_2$ is open and $F_1 \subseteq X - F_2$. Hence there exists an open set V such that $F_1 \subseteq V \subseteq \mathrm{Cl}(V) \subseteq X - F_2$. Therefore $F_1 \subseteq V$ and $F_2 \subseteq X - \mathrm{Cl}(V)$. Since $V \subseteq \mathrm{Cl}(V)$, it follows that V and $X - \mathrm{Cl}(V)$ are disjoint. This proves that X is normal. ■

The next theorem, due to Paul Urysohn (1898–1924), is a well known and important characterization of normality.

THEOREM 7.3.10 (Urysohn's Lemma) *A topological space X is normal* iff, *for any two disjoint closed sets F_1 and F_2, there exists a continuous function $f\colon X \to \mathbb{R}$, where $\mathbb{R}$ has the usual topology, such that $f(x) = 0$ for all $x \in F_1$ and $f(x) = 1$ for all $x \in F_2$.*

The proof of the necessity of the condition in Urysohn's Lemma is beyond the scope of this text and is omitted. It is left as an exercise to prove the sufficiency of the condition.

In the previous section we observed that compact Hausdorff spaces are regular. The next theorem states that these spaces are also normal.

THEOREM 7.3.11 *Every compact Hausdorff space is normal.*

Proof Let X be a compact Hausdorff space and let F_1 and F_2 be disjoint closed subsets of X. Since F_1 is closed and X is compact, it follows that F_1 is compact. Then by Lemma 6.1.18, for each $x \in F_2$, there exist disjoint open sets U_x and V_x such that $x \in U_x$ and $F_1 \subseteq V_x$. The collection $\{U_x : x \in F_2\}$ is an open cover of F_2. Since F_2 is compact, the collection has a finite subcover $\{U_{x_i} : i = 1, 2, \ldots, n\}$. Let $U = \bigcup\{U_{x_i} : i = 1, 2, \ldots, n\}$ and let $V = \bigcap\{V_{x_i} : i = 1, 2, \ldots, n\}$. Then U and V are open sets containing F_2 and F_1, respectively. Since for each $i \in \{1, 2, \ldots, n\}$, $U_{x_i} \cap V_{x_i} = \varnothing$, it follows that $U \cap V = \varnothing$. This proves that X is normal. ■

As stated in the introduction to this section, products of normal spaces are not necessarily normal and subsets of normal spaces may fail to be normal. The following example gives a non-normal product of normal spaces. Examples of normal spaces with non-normal subspaces may be found in advanced texts in topology.

Example 7.3.12

Let $X = \mathbb{R}$ have the $\mathscr{H}$ topology. Then X is normal, but the product space $X \times X$, known as the Sorgenfrey space, is not normal (see Figure 7.3.1). The basic open sets are "half-open rectangles." Let

$$A = \{(x, y) \in X \times X : y = -x\}.$$

Note that A is closed and that the relative topology for A is the discrete topology. Then consider the sets $F_1 = \{(x, y) \in A : x \text{ is rational}\}$ and

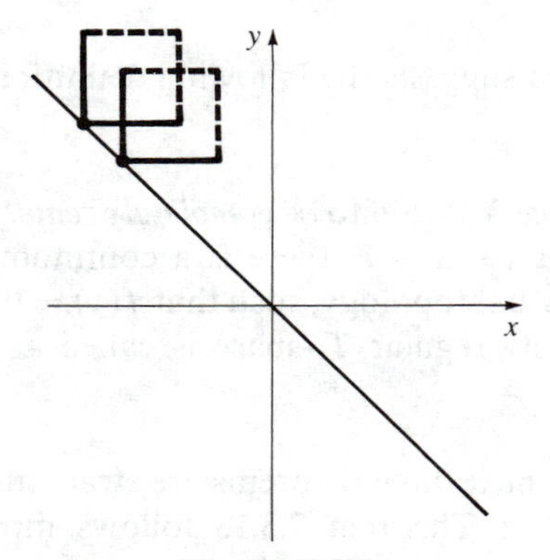

Figure 7.3.1

$F_2 = \{(x, y) \in A : x \text{ is irrational}\}$. Since F_1 and F_2 are closed subsets of A and A is a closed subset of $X \times X$, it follows that F_1 and F_2 are closed subsets of $X \times X$. It is intuitively obvious that F_1 and F_2 cannot be separated by open sets. However, the proof of this fact is quite difficult and is left for an advanced course.

There is one important product of normal spaces that is normal. If $\mathbb{R}$ has the usual topology, then $\mathbb{R}^n$ is normal for any positive integer n. The proof is similar to that of Theorem 7.3.7 except the usual distance function for $\mathbb{R}^n$ is used in place of the absolute value function.

Recall that $\mathbb{R}$ with the $\mathscr{C}$ topology is normal but not regular (see Example 7.3.2). Thus normality is not stronger than regularity. Since any indiscrete space with two or more points is normal (there are no nonempty disjoint closed sets) but not a T_0-space, normality does not even imply the T_0-property. However, if the T_1-property is combined with normality, the resulting property is stronger than regularity.

Definition 7.3.13

A normal T_1-space is called a *T_4-space.*

Note that some authors interchange the meanings of normality and the T_4-property.

The proofs of the following theorems are left as exercises.

THEOREM 7.3.14 *Every T_4-space is a T_3-space.*

THEOREM 7.3.15 *Normality is a topological property.*

THEOREM 7.3.16 *The T_4-property is a topological property.*

There are many more important separation conditions. We conclude this section by briefly mentioning some of these properties. The development of the theory of these properties is better suited to an advanced course in topology.

Urysohn's Lemma suggests the following definition.

Definition 7.3.17

A topological space X is said to be *completely regular* if, for each closed set F and each point $y \in X - F$, there is a continuous function $f: X \to \mathbb{R}$, where $\mathbb{R}$ has the usual topology, such that $f(y) = 0$ and $f(x) = 1$ for each $x \in F$. A completely regular T_1-space is called a *Tychonoff space* or a *$T_{3\frac{1}{2}}$-space.*

The proofs of the next three theorems are straightforward and are left as exercises. The proof of Theorem 7.3.18 follows directly from Urysohn's Lemma.

THEOREM 7.3.18 *Every T_4-space is a Tychonoff space.*

THEOREM 7.3.19 *Every Tychonoff space is a T_3-space.*

THEOREM 7.3.20 *The Tychonoff property is a topological property.*

Recall that a topological space is normal if any two disjoint closed sets are contained in disjoint open sets. The following definition suggests a condition stronger than normality.

Definition 7.3.21
Two nonempty subsets A and B of a topological space X are said to be *separated* provided that $A \cap \text{Cl}(B) = \varnothing$ and $\text{Cl}(A) \cap B = \varnothing$.

Definition 7.3.22
A topological space X is said to be *completely normal* if for any two separated sets A and B, there exist disjoint open sets U and V such that $A \subseteq U$ and $B \subseteq V$. A completely normal T_1-space is called a *T_5-space.*

Since nonempty disjoint closed sets are obviously separated, we have the following theorem.

THEOREM 7.3.23 *Every T_5-space is a T_4-space.*

THEOREM 7.3.24 *The T_5-property is a topological property.*
The proofs of these theorems are left as exercises.

Exercises 7.3

1. Let $X = \{a, b, c, d\}$. Determine whether each of the following spaces is regular and/or normal:

   (a) $(X, \mathcal{T})$ where $\mathcal{T} = \{X, \varnothing, \{a, b\}, \{c, d\}\}$
   (b) $(X, \mathcal{S})$ where $\mathcal{S} = \{X, \varnothing, \{a\}, \{a, b\}, \{a, b, c\}\}$
   (c) $(X, \mathcal{F})$ where $\mathcal{F} = \{X, \varnothing, \{b\}, \{c\}, \{b, c\}, \{a, b\}, \{a, b, c\}, \{b, c, d\}\}$
   (d) $(X, \mathcal{D})$ where $\mathcal{D}$ is the discrete topology
   (e) $(X, \mathcal{I})$ where $\mathcal{I}$ is the indiscrete topology.

2. Let $X = \mathbb{R}$. Determine whether each of the following spaces is regular and/or normal:

   (a) $(X, \mathcal{T})$ where $\mathcal{T} = \{X, \varnothing, (-\infty, 0), [0, +\infty)\}$
   (b) $(X, \mathcal{L})$ where $\mathcal{L} = \{X, \varnothing\} \cup \{(-\infty, x) : x \in X\}$
   (c) $(X, \mathcal{F})$ where $\mathcal{F} = \{U \subseteq X : X - U \text{ is finite or } 1 \notin U\}$
   (d) $(X, \mathcal{G})$ where $\mathcal{G} = \{U \subseteq X : 2 \in U \text{ or } U = \varnothing\}$
   (e) $(X, \mathcal{L})$ where $\mathcal{L} = \{U \subseteq X : 2 \notin U \text{ or } U = X\}$.

3. Prove that the space $(\mathbb{R}, \mathscr{H})$ is normal (Theorem 7.3.8).

4. Prove the sufficiency of the condition given in Theorem 7.3.10 (Urysohn's Lemma).

5. Prove Theorem 7.3.14.

6. Prove Theorem 7.3.15.

7. Prove Theorem 7.3.16.

8. Give a geometric argument to show that the set A in Example 7.3.12 is closed.

9. Give a geometric argument to show that the relative topology for the set A in Example 7.3.12 is the discrete topology.

10. Prove that a closed subset of a normal space is normal.

11. Let X and Y be nonempty topological spaces. Prove that if $X \times Y$ is a T_4-space, then X and Y are T_4-spaces. (Hint: Find a closed subset of $X \times Y$ that is homeomorphic to X. Then use Exercise 10 and Theorem 7.3.15.)

12. Prove Theorem 7.3.18.

13. Prove Theorem 7.3.19.

14. Prove Theorem 7.3.20.

15. Prove Theorem 7.3.23.

16. Prove Theorem 7.3.24.

17. Let A and B be nonempty disjoint subsets of a topological space X. Prove that A and B are separated iff both A and B are closed in the subspace $A \cup B$.

18. Let A and B be nonempty disjoint subsets of a topological space X. Prove that A and B are separated iff both A and B are open in the subspace $A \cup B$.

Review Exercises 7

Mark each of the following statements true or false. Briefly explain each true statement and find a counterexample for each false statement.

1. Any indiscrete topological space is not Hausdorff.

2. Any indiscrete topological space with two or more points is not Hausdorff.

3. Singleton subsets of T_2-spaces are closed.

4. Singleton subsets of T_1-spaces are closed.

5. Singleton subsets of T_0-spaces are closed.

6. Subspaces of regular spaces are regular.

7. Subspaces of T_2-spaces are T_2-spaces.

8. Closed subsets of normal spaces are normal.

9. Every normal space is regular.

10. Every normal space is a T_1-space.

11. Every T_4-space is regular.

12. Every T_4-space is a T_2-space.

13. Every T_i-space is a T_{i-1}-space for each $i \in \{1, 2, 3, 4, 5\}$.

14. If a topological space does not have any nonempty disjoint closed sets, then the space is not normal.

15. If a topological space does not have any nonempty disjoint open sets, then the space is not normal.

8

Metric Spaces

8.1 *The Metric Topology*

Recall that the distance between two points x and y on the real number line is given by $|x - y|$. This concept of distance can be used to describe the open intervals. For example, if I is an open interval with center y and radius r, then $I = \{x \in \mathbb{R} : |x - y| < r\}$. Thus the collection of all sets of the form $\{x \in \mathbb{R} : |x - y| < r\}$ forms a base for the usual topology. If an arbitrary set X has a function $d: X \times X \to \mathbb{R}$ defined on the product set $X \times X$ with properties similar to those of the distance function for the real number line, then it seems reasonable that the above process can be used to generate a topology for X. The definitions and theorems of this section make these ideas precise.

Definition 8.1.1

Let X be a set. A function $d: X \times X \to \mathbb{R}$ is said to be a *metric* for X provided that d has the following properties:

(a) $d(x, y) \geqq 0$ for all $x \in X$ and all $y \in X$.

(b) $d(x, y) = 0$ iff $x = y$ for all $x \in X$ and all $y \in X$.

(c) $d(x, y) = d(y, x)$ for all $x \in X$ and all $y \in X$.

(d) $d(x, y) \leqq d(x, z) + d(z, y)$ for all $x \in X$, all $y \in X$, and all $z \in X$ (the triangle inequality).

If d is a metric for X, then the pair (X, d) is called a *metric space.*

Properties (a)–(d) in the above definition are the natural properties for a distance function. That is, the distance between points should be nonnegative; the distance from a point to itself should be zero, and the distance between any two distinct points should be positive; the distance between two points should not depend upon the order of the points, and the triangle inequality should hold.

Example 8.1.2
If $d\colon \mathbb{R} \times \mathbb{R} \to \mathbb{R}$ is given by $d(x, y) = |x - y|$, then $(\mathbb{R}, d)$ is a metric space.

The usual distance function defined on the real number line certainly has the properties listed in Definition 8.1.1. In fact, the properties of this function are the motivation for the definition of a metric. However, there are other metrics for $\mathbb{R}$. The metric given in the next example can be defined on any nonempty set.

Example 8.1.3
Let X be any nonempty set. The function $d\colon X \times X \to \mathbb{R}$ given by

$$d(x, y) = \begin{cases} 1 & \text{if } x \neq y \\ 0 & \text{if } x = y \end{cases}$$

is a metric for X. The function d is called the discrete metric because, as we shall see later in this section, the topology associated with d is the discrete topology.

The next example gives several metrics for $\mathbb{R}^2$.

Example 8.1.4
Each of the following functions is a metric for $\mathbb{R}^2$:

(a) $d_1((x_1, y_1), (x_2, y_2)) = \sqrt{(x_1 - x_2)^2 + (y_1 - y_2)^2}$

(b) $d_2((x_1, y_1), (x_2, y_2)) = |x_1 - x_2| + |y_1 - y_2|$

(c) $d_3((x_1, y_1), (x_2, y_2)) = \max.\{|x_1 - x_2|, |y_1 - y_2|\}$.

The metric d_1 in the above example is the usual notion of distance in the plane. Each of the metrics in Example 8.1.4 can be extended in the obvious way to $\mathbb{R}^n$ for any positive integer $n \geqq 3$.

Our main purpose for introducing the concept of a metric is to study the topology generated by the metric. The following definition begins this process.

Definition 8.1.5
Let (X, d) be a metric space. For each $y \in X$ and each $r > 0$, the set $B_r(y) = \{x \in X : d(x, y) < r\}$ is called the *open ball* with center y and radius r. If there is more than one metric under consideration, the notation $B_r^d(y)$ will be used.

Example 8.1.6

Let d_1 be the metric given in Example 8.1.4. The open ball $B_1((0,0))$ is an open disk with center $(0,0)$ and radius 1. With respect to the metric d_2 in the same example, the open ball $B_1((0,0))$ is an open "diamond" with center $(0,0)$. If the metric d_3 is used, then the open ball $B_1((0,0))$ is an open "square" with sides parallel to x- and y-axes and with center $(0,0)$. (See Figures 8.1.1, 8.1.2, and 8.1.3.)

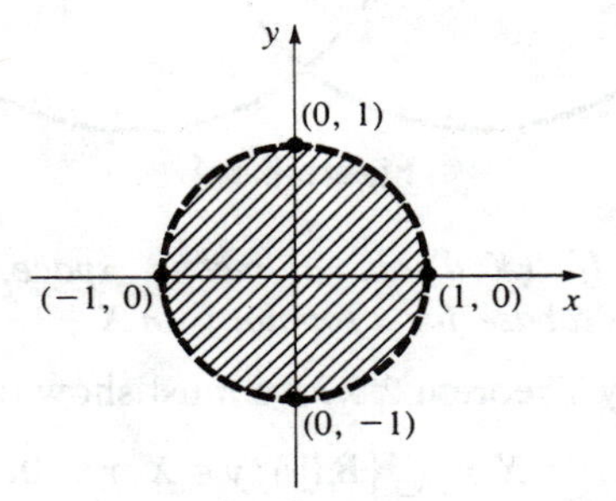

Figure 8.1.1

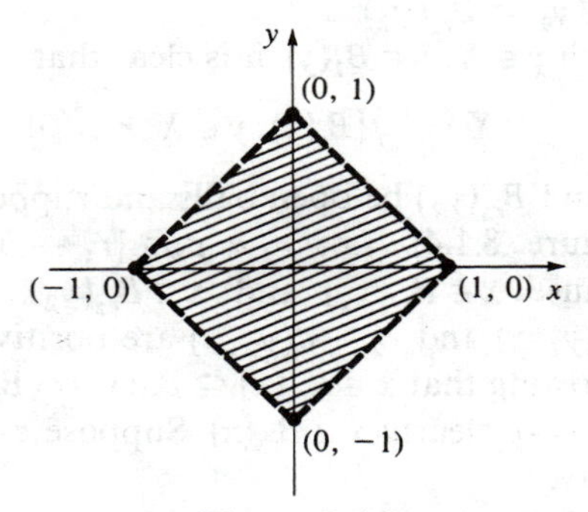

Figure 8.1.2

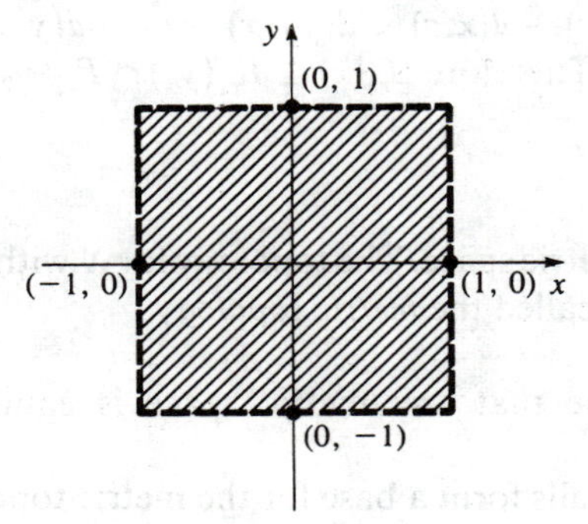

Figure 8.1.3

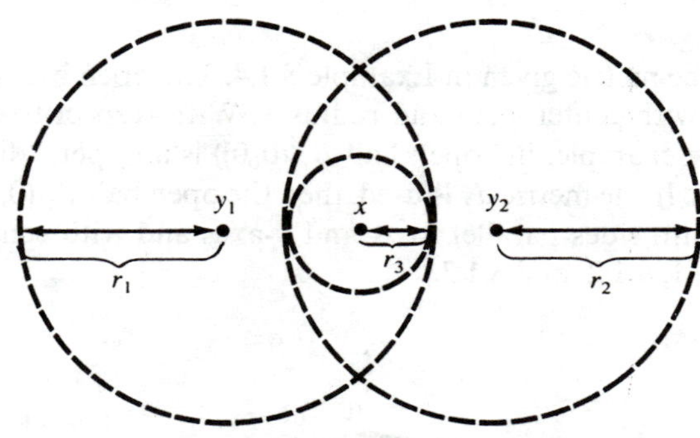

Figure 8.1.4

THEOREM 8.1.7 . *If (X, d) is a metric space, then the collection $\{B_r(y): y \in X, r > 0\}$ is a base for a topology on X.*

Proof Recall that by Theorem 2.5.7 we must show that

$$X = \bigcup\{B_r(y): y \in X, r > 0\}$$

and that for any two open balls $B_{r_1}(y_1)$ and $B_{r_2}(y_2)$ and any point $x \in B_{r_1}(y_1) \cap B_{r_2}(y_2)$, there exists an open ball $B_{r_3}(y_3)$ such that $x \in B_{r_3}(y_3) \subseteq B_{r_1}(y_1) \cap B_{r_2}(y_2)$.

Since for each $y \in X$, $y \in B_1(y)$, it is clear that

$$X = \bigcup\{B_r(y): y \in X, r > 0\}.$$

Let $B_{r_1}(y_1)$ and $B_{r_2}(y_2)$ be open balls and suppose that $x \in B_{r_1}(y_1) \cap B_{r_2}(y_2)$ (see Figure 8.1.4). Let $r_3 = \min.\{r_1 - d(y_1, x), r_2 - d(y_2, x)\}$. Note that because $x \in B_{r_1}(y_1)$ and $x \in B_{r_2}(y_2)$, obviously both the numbers $r_1 - d(y_1, x)$ and $r_2 - d(y_2, x)$ are positive. The proof will be completed by showing that $x \in B_{r_3}(x) \subseteq B_{r_1}(y_1) \cap B_{r_2}(y_2)$.

Since $d(x, x) = 0$, clearly $x \in B_{r_3}(x)$. Suppose $z \in B_{r_3}(x)$. Then by the triangle inequality,

$$d(y_1, z) \leqq d(y_1, x) + d(x, z) < d(y_1, x) + r_3 \leqq d(y_1, x) + r_1 - d(y_1, x) = r_1.$$

Thus $d(y_1, z) < r_1$ and hence $z \in B_{r_1}(y_1)$. Again by the triangle inequality, $d(y_2, z) \leqq d(y_2, x) + d(x, z) < d(y_2, x) + r_2 - d(y_2, x) = r_2$ which proves that $z \in B_{r_2}(y_2)$. Therefore $B_{r_3}(x) \subseteq B_{r_1}(y_1) \cap B_{r_2}(y_2)$ which completes the proof. ■

Definition 8.1.8

Let (X, d) be a metric space. The topology for X with the base consisting of all open balls is called the *metric topology*.

We shall assume that any metric space is equipped with its metric topology.

Since the open balls form a base for the metric topology, we know that a set is open iff it contains an open ball about each of its points. The following theorem states that we can assume the open ball is centered at the point.

THEOREM 8.1.9 *Let (X,d) be a metric space and let $U \subseteq X$. Then U is open with respect to the metric topology* iff *for each $x \in U$, there exists $r > 0$ such that $B_r(x) \subseteq U$.*

The proof is left as an exercise.

The next theorem follows from remarks made in the introduction to this section. The details of the proof are left as an exercise.

THEOREM 8.1.10 *If d is the metric on $\mathbb{R}$ given by $d(x, y) = |x - y|$ for all $x, y \in \mathbb{R}$, then the corresponding metric topology for $\mathbb{R}$ is $\mathscr{U}$.*

Henceforth we shall refer to the metric given in Theorem 8.1.10 as the usual metric for $\mathbb{R}$ and we shall assume that $(\mathbb{R}, \mathscr{U})$ is equipped with this metric. If A is a nonempty subset of $\mathbb{R}$ and d is the usual metric for $\mathbb{R}$, then d restricted to $A \times A$ is obviously a metric for A. The $\mathscr{U}$-relative topology on A is the same as the metric topology for A corresponding to the restriction of d. Thus we shall assume that $(A, \mathscr{U}_A)$ is a metric space equipped with the restriction of the usual metric for $\mathbb{R}$.

This section is concluded with two additional examples of metric topologies.

Example 8.1.11

For each of the metrics in Example 8.1.4, the metric topology is the usual topology for $\mathbb{R}^2$.

Example 8.1.12

The metric topology for the metric in Example 8.1.3 is the discrete topology.

Exercises 8.1

1. Explain why each of the following functions is *not* a metric for $\mathbb{R}$:

   (a) $d(x, y) = |x + y|$ (b) $d(x, y) = |x| - |y|$
   (c) $d(x, y) = ||x| - |y||$ (d) $d(x, y) = x^2 + y^2$
   (e) $d(x, y) = x^2 - y^2$ (f) $d(x, y) = |x^2 - y^2|$.

2. Prove that the function given in Example 8.1.3 is a metric.
3. Let $X = \mathbb{R}$ have the metric given in Example 8.1.3. Describe each of the following open balls:

   (a) $B_1(0)$ (b) $B_{1.1}(0)$ (c) $B_{1/2}(0)$.

4. Show that the metric topology for the metric in Example 8.1.3 is the discrete topology.
5. Show that the function d_3 given in Example 8.1.4 is a metric.
6. Let $\mathbb{R}^2$ have the metric d_2 given in Example 8.1.4. Sketch the open ball $B_1((1, 0))$.

7. Let $\mathbb{R}^2$ have the metric d_3 defined in Example 8.1.4. Sketch the open ball $B_1((1,0))$.
8. Prove Theorem 8.1.10.
9. Prove Theorem 8.1.9.
10. Prove that if (X, d) is a finite metric space, then the metric topology is the discrete topology.
11. Assume $\mathbb{R}$ has the usual topology and let $X = \{f: [0,1] \to \mathbb{R} : f \text{ is continuous}\}$. Show that the function $d: X \times X \to \mathbb{R}$ given by $d(f,g) = \int_0^1 |f(x) - g(x)|\,dx$ is a metric for X.
12. Let X be the set given in Exercise 11. Assume that $d: X \times X \to \mathbb{R}$ given by $d(f,g) = \max.\{|f(x) - g(x)| : x \in [0,1]\}$ is a metric for X. Describe each of the folowing open balls:
    (a) $B_1(f)$ where $f(x) = 2$ for all $x \in [0,1]$
    (b) $B_2(g)$ where $g(x) = x$ for all $x \in [0,1]$
    (c) $B_{1/2}(h)$ where $h(x) = x^2$ for all $x \in [0,1]$.

8.2 *Properties of Metric Spaces*

In this section the separation properties of metric spaces are investigated. Continuity for functions between metric spaces is also examined. Finally, subspaces of metric spaces and products of metric spaces are studied.

THEOREM 8.2.1 *Every metric space is Hausdorff.*

Proof Let (X, d) be a metric space and let x and y be two points in X. Since $x \neq y$, $d(x, y) \neq 0$. Let $r = d(x, y)/2$. We shall show that $B_r(x) \cap B_r(y) = \varnothing$. Suppose $z \in B_r(x) \cap B_r(y)$. By the triangle inequality

$$d(x, y) \leqq d(x, z) + d(z, y) < 2(d(x, y)/2) = d(x, y)$$

which is a contradiction. Thus $B_r(x)$ and $B_r(y)$ are disjoint and therefore X is Hausdorff. ■

The proof of the next theorem is analogous to the proof that $(\mathbb{R}, \mathscr{U})$ is normal.

THEOREM 8.2.2 *Every metric space is normal.*

Proof Let (X, d) be a metric space and let F_1 and F_2 be disjoint closed subsets of X. Let $x \in F_1$. Then $x \notin F_2$. Since F_2 is closed, there exists an open ball $B_{r_x}(x)$ such that $B_{r_x}(x) \cap F_2 = \varnothing$. Similarly for each $y \in F_2$, there exists an open ball $B_{r_y}(y)$ such that $B_{r_y}(y) \cap F_1 = \varnothing$. Let

$$U = \bigcup\{B_{r_x/2}(x) : x \in F_1\}$$

and let $V = \bigcup\{B_{r_y/2}(y) : y \in F_2\}$. By definition of the metric topology, both U and V are open. Obviously $F_1 \subseteq U$ and $F_2 \subseteq V$. The proof will be completed by showing that U and V are disjoint.

Suppose that $z \in U \cap V$. Since $z \in U$, there exists $x \in F_1$ such that $z \in B_{r_x/2}(x)$ and hence $d(x, z) < r_x/2$. Similarly because $z \in V$, there exists

$y \in F_2$ such that $z \in B_{r_y/2}(y)$ and thus $d(z, y) < r_y/2$. By the triangle inequality, $d(x, y) \leqq d(x, z) + d(z, y) < (r_x/2) + (r_y/2)$. Note that either $r_x \leqq r_y$ or $r_y \leqq r_x$. Suppose that $r_x \leqq r_y$. Then $d(x, y) < (r_y/2) + (r_y/2) = r_y$ and hence $x \in B_{r_y}(y)$. This is a contradiction because $B_{r_y}(y)$ is disjoint from F_1 and $x \in F_1$. Similarly, if $r_y \leqq r_x$, then $d(x, y) < r_x$ which implies that $y \in B_{r_x}(x)$. This is also a contradiction since $B_{r_x}(x)$ is disjoint from F_2 and $y \in F_2$. Thus U and V are disjoint which proves that X is normal. ■

The proofs of the following corollaries are straightforward and are left as exercises.

COROLLARY 8.2.3 *Every metric space is regular.*

COROLLARY 8.2.4 *Every metric space is a T_4-space (and hence also a T_3-space).*

The next theorem gives an "ε-δ" characterization of continuity for functions between metric spaces that is analogous to the definition of continuity given in calculus.

THEOREM 8.2.5 *Let (X, d) and (Y, e) be metric spaces. A function $f: X \to Y$ is continuous* iff, *for each $x \in X$ and each $\varepsilon > 0$, there exists $\delta > 0$ such that if $d(x, y) < \delta$, then $e(f(x), f(y)) < \varepsilon$.*

Proof ($\Rightarrow$) Assume that $f: X \to Y$ is continuous and let $x \in X$. Let $\varepsilon > 0$. Since $B^e_\varepsilon(f(x))$ is an open subset of Y, $f^{-1}(B^e_\varepsilon(f(x)))$ is an open subset of X. Because $x \in f^{-1}(B^e_\varepsilon(f(x)))$, there exists $\delta > 0$ such that $B^d_\delta(x) \subseteq f^{-1}(B^e_\varepsilon(f(x)))$. It is left as an exercise to show that if $d(x, y) < \delta$, then $e(f(x), f(y)) < \varepsilon$.

($\Leftarrow$) Assume that for each $x \in X$ and each $\varepsilon > 0$, there exists $\delta > 0$ such that, if $d(x, y) < \delta$, then $e(f(x), f(y)) < \varepsilon$. Let V be an open subset of Y. Suppose that $x \in f^{-1}(V)$. Then $f(x) \in V$. Since V is open, there exists $\varepsilon > 0$ for which $B^e_\varepsilon(f(x)) \subseteq V$. By assumption there exists $\delta > 0$ such that, if $d(x, y) < \delta$, then $e(f(x), f(y)) < \varepsilon$. We leave it as an exercise to show that $B^d_\delta(x) \subseteq f^{-1}(V)$ and hence that $f^{-1}(V)$ is open. ■

The concept of uniform continuity easily extends to functions between metric spaces.

Definition 8.2.6

Let (X, d) and (Y, e) be metric spaces. A function $f: X \to Y$ is said to be *uniformly continuous* provided that given $\varepsilon > 0$, there exists $\delta > 0$ such that for all $x_1, x_2, \in X$, $d(x_1, x_2) < \delta \Rightarrow e(f(x_1), f(x_2)) < \varepsilon$.

Recall the difference between uniform continuity and continuity. For continuous functions the value of δ may depend upon both x and ε. However, for uniformly continuous functions the choice of δ depends only upon the value of ε.

The proof of the next theorem is exactly analogous to that of Theorem 6.2.9 and is left as an exercise.

THEOREM 8.2.7 *Let (X, d) and (Y, e) be metric spaces and let $f: X \to Y$ be a continuous function. If X is compact, then f is uniformly continuous.*

One implication of the Heine–Borel Theorem (Theorem 6.1.20) extends to metric spaces.

Definition 8.2.8

A subset A of a metric space (X, d) is said to be *bounded* provided that there exist $x \in X$ and $r > 0$ such that $A \subseteq B_r(x)$.

THEOREM 8.2.9 *Let (X, d) be a metric space. If A is a compact subset of X, then A is closed and bounded.*

Proof Let A be a compact subset of X. Since by Theorem 8.2.1 X is Hausdorff, it follows that A is closed (Theorem 6.1.19). To see that A is bounded, suppose that $x \in A$. The collection $\{B_r(x): r > 0\}$ is an open cover of A. Since A is compact, the collection has a finite subcover

$$\{B_{r_i}(x): i = 1, 2, \ldots, n\}.$$

If $r_0 = \max.\{r_1, r_2, \ldots, r_n\}$, then clearly $A \subseteq B_{r_0}(x)$. ■

The next example shows that the other implication of the Heine–Borel Theorem does not extend to metric spaces.

Example 8.2.10

Let $\mathbb{R}$ have the metric $d: X \times X \to \mathbb{R}$ given by

$$d(x, y) = \begin{cases} 1 & \text{if } x \neq y \\ 0 & \text{if } x = y \end{cases}$$

Since the metric topology for $\mathbb{R}$ is the discrete topology, the set $[0, 1]$ is closed and bounded but not compact. In fact, $\mathbb{R}$ is closed and bounded.

Example 8.2.10 also shows that the Bolzano–Weierstrass Theorem (Theorem 6.1.22) does not extend to metric spaces. The set $[0, 1]$ is bounded and infinite but has no limit point.

If X is any nonempty set, then there exists a metric for X. For example, the metric in Example 8.2.10 can be defined for X. However, if X is a topological space, then there may or may not be a metric for X such that the metric topology is the same as the original topology on X.

Definition 8.2.11

A topological space $(X, \mathcal{T})$ is said to be *metrizable* provided that there exists a metric d for X such that the metric topology is the same as $\mathcal{T}$.

Example 8.2.12

Any finite nondiscrete topological space is not metrizable (see Exercise 10 in Section 8.1).

Much research has been done in the area of metrizability. Several well known theorems giving sufficient and necessary conditions for a space to be metrizable can be found in advanced texts in topology.

THEOREM 8.2.13 *Metrizability is a topological property.*

Proof Let X and Y be homeomorphic topological spaces. Assume that X is metrizable and let d be a metric for X. Let $f: X \to Y$ be a homeomorphism. Define the function $e: Y \times Y \to \mathbb{R}$ by $e(x, y) = d(f^{-1}(x), f^{-1}(y))$. It is left as an exercise to show that e is a metric for Y. Let $\mathscr{T}$ be the original topology on Y and $\mathscr{T}_e$ denote the metric topology on Y induced by e. We shall show that $\mathscr{T} = \mathscr{T}_e$.

Let $U \in \mathscr{T}$ and suppose that $y \in U$. Since $f^{-1}(U)$ is an open subset of X and $f^{-1}(y) \in f^{-1}(U)$, there exists $r > 0$ such that $B_r^d(f^{-1}(y)) \subseteq f^{-1}(U)$. In order to show that $B_r^e(y) \subseteq U$, let $z \in B_r^e(y)$. Then

$$d(f^{-1}(y), f^{-1}(z)) = e(y, z) < r.$$

Thus $f^{-1}(z) \in B_r^d(f^{-1}(y)) \subseteq f^{-1}(U)$ and hence $z \in U$. This proves that $B_r^e(y) \subseteq U$ and hence that $U \in \mathscr{T}_e$. Thus $\mathscr{T} \subseteq \mathscr{T}_e$.

Let $V \in \mathscr{T}_e$. In order to show that $V \in \mathscr{T}$, we shall show that $f^{-1}(V)$ is an open subset of X. Let $x \in f^{-1}(V)$. Then $f(x) \in V$. Since $V \in \mathscr{T}_e$, there exists $r > 0$ such that $B_r^e(f(x)) \subseteq V$. In order to show that $B_r^d(x) \subseteq f^{-1}(V)$, suppose $w \in B_r^d(x)$. Then

$$r > d(x, w) = d(f^{-1}(f(x)), f^{-1}(f(w))) = e(f(x), f(w)).$$

Therefore $f(w) \in B_r^e(f(x)) \subseteq V$ and hence $w \in f^{-1}(V)$. This proves that $B_r^d(x) \subseteq f^{-1}(V)$ and hence that $f^{-1}(V)$ is an open subset of X. Because f is a homeomorphism, $V = f(f^{-1}(V))$ is a $\mathscr{T}$-open subset of Y. Therefore $V \in \mathscr{T}$ and $\mathscr{T}_e \subseteq \mathscr{T}$.

This completes the proof that $\mathscr{T} = \mathscr{T}_e$. ■

Subsets of metric spaces and products of metric spaces are investigated next.

Suppose $(X, \mathscr{T})$ is a metrizable space and A is a subset of X. Then there is a metric $d: X \times X \to \mathbb{R}$ for the set X such that $\mathscr{T}$ is the same as the metric topology induced by d. It is not difficult to show that the function d restricted to $A \times A$ is a metric for the set A. It can then be shown that the metric topology on A induced by the restriction of d is the same as the relative topology on A. Hence we have the following theorem.

THEOREM 8.2.14 *If the space $(X, \mathscr{T})$ is metrizable and A is a subset of X, then $(A, \mathscr{T}_A)$ is metrizable.*

The details of the proof are left as an exercise.

THEOREM 8.2.15 *Let X and Y be nonempty topological spaces. The product space $X \times Y$ is metrizable* iff *both X and Y are metrizable.*

Proof ($\Leftarrow$) Assume that X and Y are metrizable. Let d be the metric for X and e the metric for Y. Define the function $m: (X \times Y) \times (X \times Y) \to \mathbb{R}$ by $m((x, y), (z, w)) = d(x, z) + e(y, w)$. It is left as an exercise to show that m is a metric for the set $X \times Y$. Let $\mathscr{T}_p$ denote the product topology on $X \times Y$ and denote the metric topology on $X \times Y$ with respect to m by $\mathscr{T}_m$. We must show that $\mathscr{T}_p = \mathscr{T}_m$.

Suppose $U \in \mathscr{T}_p$. In order to show that $U \in \mathscr{T}_m$, we must show that, for each $(x, y) \in U$, there exists $r > 0$ such that $B_r^m((x, y)) \subseteq U$. Let $(x, y) \in U$. Since $U \in \mathscr{T}_p$, there exist open balls $B_a^d(x)$ and $B_b^e(y)$ such that $(x, y) \in B_a^d(x) \times B_b^e(y) \subseteq U$. Let $r = \min.\{a, b\}$. We shall show that $B_r^m((x, y)) \subseteq B_a^d(x) \times B_b^e(x)$. Suppose that $(z, w) \in B_r^m((x, y))$. Then

$$m((z, w), (x, y)) = d(z, x) + e(w, y) < r.$$

Since $r = \min.\{a, b\}$, it follows that $d(z, x) < a$ and $e(w, y) < b$. Thus $z \in B_a^d(x)$ and $w \in B_b^e(y)$ and hence $(z, w) \in B_a^d(x) \times B_b^e(y)$. Therefore

$$B_r^m((x, y)) \subseteq B_a^d(x) \times B_b^e(y) \subseteq U$$

and thus $U \in \mathscr{T}_m$. This proves that $\mathscr{T}_p \subseteq \mathscr{T}_m$.

Let $V \in \mathscr{T}_m$. In order to show that $V \in \mathscr{T}_p$, we shall show that, for each $(x, y) \in V$, there exist open balls $B_a^d(x)$ and $B_b^e(y)$ for which $(x, y) \in B_a^d(x) \times B_b^e(y) \subseteq V$. Let $(x, y) \in V$. Since $V \in \mathscr{T}_m$, there exists $r > 0$ such that $(x, y) \in B_r^m((x, y)) \subseteq V$. We shall show that $B_{r/2}^d(x) \times B_{r/2}^e(y) \subseteq B_r^m((x, y))$. Suppose $(z, w) \in B_{r/2}^d(x) \times B_{r/2}^e(y)$. Then $d(x, z) < r/2$ and $e(y, w) < r/2$ and hence $m((x, y), (z, w)) = d(x, z) + e(y, w) < (r/2) + (r/2) = r$. Therefore $(z, w) \in B_r^m((x, y))$ and thus $B_{r/2}^d(x) \times B_{r/2}^e(y) \subseteq B_r^m(x, y)) \subseteq V$. Hence $V \in \mathscr{T}_p$ and therefore $\mathscr{T}_m \subseteq \mathscr{T}_p$.

The proof that X and Y are metrizable if $X \times Y$ is metrizable is left as an exercise. ■

An induction argument will prove the following corollary.

COROLLARY 8.2.16 *If $X_1, X_2, \ldots, X_n$ are nonempty topological spaces, then the product space $X_1 \times X_2 \times \cdots \times X_n$ is metrizable* iff *X_i is metrizable for each $i \in \{1, 2, \ldots, n\}$.*

Exercise 8.2

1. Give a direct proof, without using Theorem 8.2.1, that every metric space is a T_1-space.
2. Give a direct proof, without using Theorem 8.2.2, that every metric space is regular.
3. Use Theorem 8.2.1 and Theorem 8.2.2 to prove Corollaries 8.2.3 and 8.2.4.
4. Complete the proof of Theorem 8.2.5.

5. Let (X, d) and (Y, e) be metric spaces. Prove that a function $f\colon X \to Y$ is continuous iff, for each $x \in X$ and each $\varepsilon > 0$, there exists $\delta > 0$ such that $f(B^d_\delta(x)) \subseteq B^e_\varepsilon(f(x))$.
6. Complete the proof of Theorem 8.2.13.
7. Prove Theorem 8.2.14.
8. Complete the proof of Theorem 8.2.15.
9. Let (X, d) be a metric space. Show that the function $e\colon X \times X \to \mathbb{R}$ given by $e(x, y) = \min.\{1, d(x, y)\}$ is a metric for X.
10. Show that the metric topology induced by the metric e given in Exercise 9 is the same as the metric topology induced by d.
11. Let (X, d) be a metric space. Let $y \in X$ and let $r > 0$. Show that the set $\{x \in X : d(x, y) \leqq r\}$ is closed. (This set is called the *closed ball* with center y and radius r and is denoted by $B_r^-(y)$.)
12. Let $X = \mathbb{R}$ have the metric e given by $e(x, y) = \min.\{1, |x - y|\}$ for all x, $y \in \mathbb{R}$ (see Exercise 9). Find $B_1^-(0)$ and $\mathrm{Cl}(B_1(0))$ (see Exercise 11).
13. Let (X, d) be a metric space. Prove that $d\colon X \times X \to \mathbb{R}$, where $X \times X$ has the product topology and $\mathbb{R}$ has the usual topology, is continuous.
14. Prove Theorem 8.2.7.

8.3 *Sequences*

The concept of a sequence as developed in elementary calculus extends to a general topological space. Although sequences will be defined for general topological spaces, we shall see that sequences play a very important role in metric spaces.

Definition 8.3.1

A *sequence* in a set X is a function $s\colon \mathbb{Z}^+ \to X$, where $\mathbb{Z}^+$ is the set of positive integers. For each $n = \mathbb{Z}^+$, $s(n)$ will be denoted by s_n and will be referred to as the "nth term" of the sequence s.

Example 8.3.2

Let $X = \mathbb{R}$. If $s_n = 1/(n + 1)$ for each $n \in \mathbb{Z}^+$, then s is a sequence in $\mathbb{R}$.

Example 8.3.3

Let $X = \mathbb{R}$. The function s given by $s_n = n/(n + 1)$ for each $n \in \mathbb{Z}^+$ is a sequence in $\mathbb{R}$.

Recall that the theory of convergence of sequences is very important in calculus. The following statement defines convergence of a sequence in a general topological space. Convergence of sequences is somewhat more complicated in a general topological space than on the real number line. For example, there are topological spaces in which a sequence can converge to more than one point.

Definition 8.3.4

Let X be a topological space and let $x \in X$. A sequence s in X is said to *converge to* x provided that, for any open set U containing x, there exists a positive integer N such that if $n > N$ then $s_n \in U$. The convergence of a sequence s to a point x is denoted by $s \to x$ or by $s_n \to x$. If a sequence s converges to x, then x is said to be a *limit* of s.

Example 8.3.5

The sequence s in Example 8.3.2 converges to 0 in the space $(\mathbb{R}, \mathscr{U})$. To see this, let U be any $\mathscr{U}$-open set containing 0. There exists $\varepsilon > 0$ such that $0 \in (-\varepsilon, \varepsilon) \subseteq U$. We must find a positive integer N such that, if $n > N$, then $s_n = 1/(n+1) < \varepsilon$. That is, $1/\varepsilon < n + 1$ or $n > (1/\varepsilon) - 1$. Hence, if N is any positive integer larger than $(1/\varepsilon) - 1$, then $n > N$ implies that $1/(n+1) \in (-\varepsilon, \varepsilon) \subseteq U$.

Example 8.3.6

Let $X = \mathbb{R}$ have the $\mathscr{C}$ topology. If s is the sequence in X defined by $s_n = n$ for each $n \in \mathbb{Z}^+$, then s converges to each point in X. In order to prove this, let $x \in X$ and let U be any open set containing x. Then $U = (a, +\infty)$ for some $a \in X$. If N is a positive integer such that $N > a$, then $n > N$ implies that $s_n = n \in U$.

The next theorem is useful in showing that a sequence does not converge to a given point.

THEOREM 8.3.7 *Let X be a topological space and let $x \in X$. A sequence s in X does not converge to x iff there exists an open set U containing x such that, for any positive integer N, there is an integer $n > N$ for which $s_n \notin U$.*

The proof is left as an exercise.

Example 8.3.8

Let $X = \mathbb{R}$ have the $\mathscr{C}$ topology. If s is the sequence in X given by $s_n = -n$ for each $n \in \mathbb{Z}^+$, then s does not converge to any point in X.

As we observed in Example 8.3.6, it is possible for a sequence to converge to more than one point. This is obviously not a desirable property for a sequence. The following theorem gives a condition that will eliminate this problem.

THEOREM 8.3.9 *If X is a Hausdorff space, then any sequence in X either converges to a unique point in X or does not converge to any point in X.*

Proof Let X be a Hausdorff space and let s be a sequence in X. Suppose that s converges to two distinct points x and y in X. Since X is Hausdorff, there exist disjoint open sets U and V containing x and y, respectively. The convergence of s to x implies that there exists a positive integer N_1 such that, if $n > N_1$, then $s_n \in U$. Similarly, since s converges to y, there exists a positive integer N_2 such that $n > N_2$ implies that $s_n \in V$. Let $N = \max.\{N_1, N_2\}$. Then for $n > N$, we have that $s_n \in U \cap V$ which is a contradiction since U and V are disjoint. Therefore s either converges to exactly one point or does not converge to any point in X. ■

COROLLARY 8.3.10 *Every sequence in a metric space either converges to exactly one point or does not converge to any point in the space.*

Suppose that X and Y are topological spaces and $f: X \to Y$ is a function. If s is a sequence in X, then $f \circ s$ is a sequence in Y. The next theorem relates the convergence of $f \circ s$ to the convergence of s.

THEOREM 8.3.11 *Let X and Y be topological spaces with $x \in X$ and let $f: X \to Y$ be a continuous function. If s is a sequence in X which converges to x, then $f \circ s$ is a sequence in Y which converges to $f(x)$.*

Proof Let U be any open subset of Y containing $f(x)$. Then $f^{-1}(U)$ is an open subset of X containing x. Since $s \to x$, there exists a positive integer N such that, if $n > N$, then $s_n \in f^{-1}(U)$. However, if $s_n \in f^{-1}(U)$, then $f \circ s(n) = f(s(n)) = f(s_n) \in U$. This proves that $f \circ s \to f(x)$. ■

The next theorem gives a relationship between sequences and limit points.

THEOREM 8.3.12 *Let X be a topological space with $A \subseteq X$ and $x \in X$. If there is a sequence in $A - \{x\}$ which converges to x, then x is a limit point of A.*

The proof is left as an exercise.

The following example shows that the converse of Theorem 8.3.12 does not hold.

Example 8.3.13

Let $X = \mathbb{R}$ have the topology $\mathcal{T} = \{U \subseteq X : U = \varnothing$ or $X - U$ is countable$\}$. Let $x \in X$ and let $A = X - \{x\}$. Then x is a limit point of A, but there is no sequence in A which converges to x. We leave it as an exercise to show that x is a limit point of A. To see that no sequence in A converges to x, note that if s is any sequence in A, then $U = X - \{s_n : n \in \mathbb{Z}^+\}$ is an open set containing x and that obviously $s_n \notin U$ for each $n \in \mathbb{Z}^+$.

This last theorem of this section characterizes the convergence of sequences in metric spaces. Note the similarity between the characterization and the definition of convergence given in elementary calculus. The proof is left as an exercise.

THEOREM 8.3.14 *Let (X, d) be a metric space and let $x \in X$. A sequence s in X converges to x* iff, *given* $\varepsilon > 0$, *there exists a positive integer N such that, if $n > N$, then $d(s_n, x) < \varepsilon$.*

Exercises 8.3

1. Show that the sequence in Example 8.3.3 converges to 1 in $(\mathbb{R}, \mathscr{U})$ but does not converge to 1 in $(\mathbb{R}, \mathscr{H})$.
2. Let $X = \mathbb{R}$ have the $\mathscr{C}$ topology. Prove that if s is the sequence in X given by $s_n = 1/n$ for each $n \in \mathbb{Z}^+$ then for each $x \leqq 0$, $s_n \to x$.
3. Let $X = \mathbb{R}$ have the $\mathscr{C}$ topology. Prove that a sequence s in X converges iff the set $\{s_n : n \in \mathbb{Z}^+\}$ has a lower bound.
4. Let $X = \mathbb{R}$ have the topology $\mathscr{T} = \{U \subseteq X : 1 \in U \text{ or } U = \varnothing\}$. Let s be the sequence in X given by $s_n = 1$ for each $n \in \mathbb{Z}^+$. Does s converge? If it does, find all numbers to which it converges.
5. Let $X = \mathbb{R}$ have the topology $\mathscr{T} = \{U \subseteq X : 1 \notin U \text{ or } U = X\}$. Let s be the sequence in X given by $s_n = 1$ for each $n \in \mathbb{Z}^+$. Does s converge? If it converges, find all limits of s.
6. Prove Theorem 8.3.7.
7. Show that the sequence in Example 8.3.8 does not converge to any point in X.
8. Prove that if X is an indiscrete space then every sequence in X converges to each point in X.
9. Characterize the convergent sequences in a discrete topological space.
10. Prove Corollary 8.3.10.
11. Prove Theorem 8.3.12.
12. In Example 8.3.13 show that x is a limit point of the set A.
13. Prove Theorem 8.3.14.
14. Let (X, d) be a metric space. Prove that if s is a convergent sequence in X, then the set $\{s_n : n \in \mathbb{Z}^+\}$ is a bounded subset of X.
15. Let (X, d) be a metric space and let $x \in X$. Show that a sequence s in X converges to x iff for each open ball $B_r(x)$, there exists a positive integer N such that if $n > N$ then $s_n \in B_r(x)$.
16. Let X be a topological space with $x \in X$. Prove that a sequence s in X converges to x iff for each basic open set U containing x there exists a positive integer N such that, if $n > N$, then $s_n \in U$.

8.4 *Complete Metric Spaces*

If a sequence in a metric space converges, then the terms of the sequence become close to the limit and hence close to each other. It seems reasonable that the converse of this statement should hold. That is, if the terms of a sequence become close to each other, then the sequence should converge. However, there are metric spaces in which this does not hold. For example, consider the metric space $X = (0, 1]$ with the restriction of the usual metric for $\mathbb{R}$ and the sequence s in X given by $s_n = 1/n$ for each $n \in \mathbb{Z}^+$. To see that the terms of s become close to each other, assume N, m, $n \in \mathbb{Z}^+$ and that m

and n are larger than N. Then

$$|s_m - s_n| = |(1/m) - (1/n)| \leqq |1/m| + |1/n| < (1/N) + (1/N) = 2/N.$$

Therefore by choosing N sufficiently large, we can make $|s_m - s_n|$ arbitrarily small. However, the sequence s does not converge in X. (If s converged in X, then as a sequence in $(\mathbb{R}, \mathscr{U})$ it would have two limits and this would contradict the fact that $(\mathbb{R}, \mathscr{U})$ is Hausdorff.)

Sequences with the property that the terms become arbitrarily close to each other may not converge, but these sequences are clearly different from other divergent sequences and play an important role in the theory of metric spaces. The following statement is the formal definition of this type of sequence.

Definition 8.4.1

Let (X, d) be a metric space. A sequence s in X is said to be a *Cauchy sequence* if for every $\varepsilon > 0$, there is a positive integer N such that, if $m \geqq N$ and $n \geqq N$, then $d(s_m, s_n) < \varepsilon$.

Our first theorem in this section states that the collection of all Cauchy sequences in a space contains the convergent sequences.

THEOREM 8.4.2 *Let (X, d) be a metric space. Every convergent sequence in X is also a Cauchy sequence.*

Proof Let s be a convergent sequence in X. Assume that $s \to x$. Let $\varepsilon > 0$. By Theorem 8.3.14 there exists a positive integer N for which $n > N$ implies that $d(s_n, x) < \varepsilon/2$. Assume that $m > N$ and $n > N$. Then by the triangle inequality, $d(s_m, s_n) \leqq d(s_m, x) + d(x, s_n) < (\varepsilon/2) + (\varepsilon/2) = \varepsilon$. This proves that s is a Cauchy sequence. ■

From the discussion at the beginning of this section, it follows that the sequence s in $X = (0, 1]$ given by $s_n = 1/n$ for each $n \in \mathbb{Z}^+$ is a Cauchy sequence which does not converge in X. Therefore the converse of Theorem 8.4.2 does not hold.

Definition 8.4.3

A metric space (X, d) is said to be *complete* if every Cauchy sequence in X converges to a point in X.

Obviously the space $X = (0, 1]$ with the restriction of the usual metric for $\mathbb{R}$ is not complete. Note that X is not compact.

The following theorem gives a condition on a metric space that is sufficient for completeness.

THEOREM 8.4.4 *Every compact metric space is complete.*

Proof Let (X, d) be a compact metric space and let s be a Cauchy sequence in X. Suppose that s does not converge in X. Then for each $x \in X$, there exists an open ball $B_{r_x}(x)$ containing x such that, for any positive integer N, there exists $n > N$ for which $s_n \notin B_{r_x}(x)$. For each $x \in X$, let $U_x = B_{r_x/2}(x)$. The collection $\{U_x : x \in X\}$ is an open cover of X. Since X is compact, this collection has a finite subcover $\{U_{x_i} : i = 1, 2, \ldots, k\}$. Let $r = \min.\{r_{x_1}, r_{x_2}, \ldots, r_{x_k}\}$. There exists a positive integer N such that, if $m \geqq N$ and $n \geqq N$, then $d(s_m, s_n) < r/2$. Let $i \in \{1, 2, \ldots, k\}$ such that $s_N \in U_{x_i}$. Then $d(s_N, x_i) < r_{x_i}/2$. Now if $m > N$, then by the triangle inequality, $d(s_m, x_i) \leqq d(s_m, s_N) + d(s_N, x_i) < (r/2) + (r_{x_i}/2) \leqq (r_{x_i}/2) + (r_{x_i}/2) = r_{x_i}$. Thus if $m > N$, then $s_m \in B_{r_{x_i}}(x_i)$. This contradicts the choice of the open ball $B_{r_{x_i}}(x_i)$. Therefore s must converge in X and hence (X, d) is a complete metric space. ■

Our next goal is to prove that the space $(\mathbb{R}, \mathscr{U})$ is complete. The following definition and lemma are useful.

Definition 8.4.5

A sequence s in a metric space (X, d) is said to be *bounded* if the set $\{s_n : n \in \mathbb{Z}^+\}$ is a bounded subset of X.

LEMMA 8.4.6 *Every Cauchy sequence is bounded.*

Proof Let (X, d) be a metric space and let s be a Cauchy sequence in X. There exists a positive integer N such that, if $m \geqq N$ and $n \geqq N$, then $d(s_m, s_n) < 1$. Then for each $m \geqq N$, $d(s_m, s_N) < 1$. Let

$$K = \max.\{d(s_i, s_N) : i = 1, 2, \ldots, N - 1\}.$$

Now if $L = \max.\{K, 1\}$, then for each positive integer m, $d(s_m, s_N) < L$ and hence the sequence s is bounded. ■

THEOREM 8.4.7 *If d is the usual metric for $\mathbb{R}$, then $(\mathbb{R}, d)$ is a complete metric space.*

Proof Let s be a Cauchy sequence in $(\mathbb{R}, d)$. By Lemma 8.4.6 the sequence s is bounded. Therefore there is a closed interval $[a, b]$ for which

$$\{s_n : n \in \mathbb{Z}^+\} \subseteq [a, b].$$

Since s is a Cauchy sequence in $\mathbb{R}$, clearly s is a Cauchy sequence in $[a, b]$ (see Exercise 4). Since the corresponding metric topology for $\mathbb{R}$ is $\mathscr{U}$, the Heine–Borel Theorem implies that $[a, b]$ is compact. Then by Theorem 8.4.4, $[a, b]$ is a complete metric space. Thus the sequence s converges in $[a, b]$. Obviously s also converges in $\mathbb{R}$. This proves that $(\mathbb{R}, d)$ is complete. ■

Example 8.4.8

If $\mathbb{Q}$ is equipped with the restriction of the usual metric for $\mathbb{R}$, then the metric space $\mathbb{Q}$ is not complete. To see this, assume that $\mathbb{R}$ has its usual metric and let s be any sequence of rational numbers converging to $\sqrt{2}$. Then s is a sequence in $\mathbb{R}$ as well as $\mathbb{Q}$. As a sequence in $\mathbb{R}$, s converges and hence is a Cauchy sequence in $\mathbb{R}$. Clearly s is also a Cauchy sequence in $\mathbb{Q}$. However, s does not converge in $\mathbb{Q}$.

The fact that (with respect to the usual metric) the real numbers are complete and the rational numbers are not complete is a major reason for the importance of the completeness property. In fact, the completeness property is often used as a means of differentiating between the two spaces.

In a future course it will be shown that every metric space is a dense subspace of a complete metric space, called its completion. In particular, it will be proved that the completion of the rational numbers is the real numbers.

Exercises 8.4

1. Which of the following sequences are Cauchy sequences in $(\mathbb{R}, \mathscr{U})$?

(a) $s_n = 1 - (1/n)$ **(b)** $s_n = n^2$ **(c)** $s_n = (n + 1)/n$
(d) $s_n = (n^2 + 1)/n$ **(e)** $s_n = \cos(n\pi)$

2. Assume that each of the following subsets of $\mathbb{R}$ has the $\mathscr{U}$-relative topology. Which of the spaces are complete? If a space is not complete, find a Cauchy sequence that does not converge.

(a) $(0, 1)$ **(b)** $[0, 1]$ **(c)** $[0, 1)$
(d) $[0, 2] \cup [3, 5] \cup \{6\}$ **(e)** $\mathbb{R} - \{0\}$ **(f)** $\mathbb{R}$
(g) $\mathbb{Q}$ **(h)** $\mathbb{Q} \cap [0, 4]$

3. Show that completeness is not a topological property. Also show that a subspace of a complete space is not necessarily complete.

4. Let (X, d) be a metric space and let $A \subseteq X$. Let d_A denote the metric d restricted to $A \times A$. Show that if s is a Cauchy sequence in (X, d) and $\{s_n : n \in \mathbb{Z}^+\} \subseteq A$, then s is a Cauchy sequence in (A, d_A). (This result was used in the proof of Theorem 8.4.7.)

5. Let (X, d) and (Y, e) be metric spaces and let $f: X \to Y$ be a continuous function. Prove that if X is compact and s is a Cauchy sequence in (X, d), then $f \circ s$ is a convergent sequence in (Y, e).

6. Let (X, d) and (Y, e) be metric spaces and let $f: X \to Y$ be a uniformly continuous function. Prove that if s is a Cauchy sequence in X then $f \circ s$ is a Cauchy sequence in Y.

7. Show that Exercise 6 is false if uniform continuity is replaced by continuity. (Consider $(0, 1]$ with the $\mathscr{U}$-relative topology, the sequence s in $(0, 1]$ given by $s_n = 1/n$ for each $n \in \mathbb{Z}^+$, and the function $f: X \to \mathbb{R}$, where $\mathbb{R}$ has the usual metric, given by $f(x) = 1/x$.)

8. Let X be a nonempty set and let d be the metric on X given by

$$d(x, y) = \begin{cases} 1 & \text{if } x \neq y \\ 0 & \text{if } x = y \end{cases}$$

for all $x, y \in X$. Describe the Cauchy sequences in (X, d). Determine if (X, d) is complete.

Review Exercises 8

Mark each of the following statements true or false. Briefly explain each true statement and find a counterexample for each false statement.

1. If X is any set, then there is a metric d for X.
2. If $(X, \mathscr{T})$ is any topological space, then there is a metric d such that the metric topology induced by d is the same as $\mathscr{T}$.
3. Every open ball in a metric space is open with respect to the metric topology.
4. Every open subset of a metric space (with respect to the metric topology) is an open ball.
5. If there are two different metrics for a given set, then the corresponding metric topologies are also different.
6. If (X, d) is any nonempty metric space, then for any positive integer N there exist points x and y in X such that $d(x, y) > N$.
7. If (X, d) is any nonempty metric space, then for any positive integer N there exist distinct points x and y in X such that $d(x, y) < 1/N$.
8. The intersection of two open balls is an open ball.
9. The intersection of two open balls contains an open ball about each of its points.
10. If X is a discrete topological space, then X is metrizable.
11. If X is an indiscrete topological space, then X is metrizable.
12. Every compact metric space is complete.
13. Every complete metric space is compact.
14. Every Cauchy sequence is a convergent sequence.
15. Every convergent sequence (in a metric space) is a Cauchy sequence.
16. Every complete metric space is normal.
17. Every metric space is normal.
18. Every homeomorphic image of a metric space is a metric space.
19. Every homeomorphic image of a complete metric space is a complete metric space.
20. If (X, d) and (Y, e) are metric spaces and $f: X \to Y$ is a homeomorphism, then $d(x, y) = e(f(x), f(y))$ for all $x, y \in X$.

Bibliography

Arnold, B. H. *Intuitive Concepts in Elementary Topology*. Englewood Cliffs, NJ: Prentice-Hall, 1962.

Baum, J. D. *Elements of Point-Set Topology*. Englewood Cliffs, NJ: Prentice-Hall, 1964.

Croom, F. H. *Principles of Topology*. New York: Saunders, 1989.

Dixmier, J. *General Topology*. New York: Springer-Verlag, 1984.

Dugundji, J. *Topology*. Boston: Allyn and Bacon, 1966.

Gamelin, T. W. and Greene, R. E. *Introduction to Topology*. New York: Saunders, 1983.

Gemignani, M. C. *Elementary Topology*, 2nd ed. Reading, MA: Addison-Wesley, 1967.

Greever, J. *Theory and Examples of Point-set Topology*. Belmont, CA: Brooks-Cole, 1967.

Kelley, J. L. *General Topology*. New York: Van-Nostrand, 1955.

Long, P. E. *An Introduction to General Topology*. Columbus, OH: Charles E. Merrill, 1971.

Manheim, J. H. *The Genesis of Point Set Topology*. Ann Arbor, MI: University Microfilms, 1962.

Mansfield, M. J. *Introduction to Topology*. Huntington, NY: Robert E. Krieger, 1972.

Mendelson, B. *Introduction to Topology*. Boston: Allyn and Bacon, 1968.

Moore, T. O. *Elementary General Topology*. Englewood Cliffs, NJ: Prentice-Hall, 1964.

Munkres, J. R. *Topology: A First Course*. Englewood Cliffs, NJ: Prentice-Hall, 1975.

Smith, D., Eggen, M., and St. Andre, R. *A Transition to Advanced Mathematics*, 2nd ed. Belmont, CA: Brooks-Cole, 1986.

Steen, L. A. and Seebach, J. A. *Counterexamples in Topology*. New York: Holt, Rinehart, and Winston, 1970.

Wilansky, A. *Topology for Analysis*. Waltham, MA: Ginn, 1970.

Willard, S. *General Topology*. Reading, MA: Addison-Wesley, 1970.

Index